Society of Earth Scientists Series

Series editor

Satish C. Tripathi, Lucknow, India

For further volumes:
http://www.springer.com/series/8785

The Society of Earth Scientists Series aims to publish selected conference proceedings, monographs, edited topical books/text books by leading scientists and experts in the field of geophysics, geology, atmospheric and environmental science, meteorology and oceanography as Special Publications of The Society of Earth Scientists. The objective is to highlight recent multidisciplinary scientific research and to strengthen the scientific literature related to Earth Sciences. Quality scientific contributions from all across the Globe are invited for publication under this series.

Prashant K. Srivastava · Saumitra Mukherjee
Manika Gupta · Tanvir Islam
Editors

Remote Sensing Applications in Environmental Research

Springer

Editors
Prashant K. Srivastava
Hydrological Sciences
ESSIC/NASA GSFC
Greenbelt, MD
USA

Manika Gupta
Civil Engineering
Indian Institute of Technology (IIT)
New Delhi
India

Saumitra Mukherjee
School of Environmental Sciences
Jawaharlal Nehru University
New Delhi
India

Tanvir Islam
Atmospheric Sciences
CIRA/NOAA
College Park, MD
USA

ISSN 2194-9204 ISSN 2194-9212 (electronic)
ISBN 978-3-319-35941-0 ISBN 978-3-319-05906-8 (eBook)
DOI 10.1007/978-3-319-05906-8
Springer Cham Heidelberg New York Dordrecht London

Softcover reprint of the hardcover 1st edition 2014

Printed on acid-free paper

Springer is part of Springer Science+Business Media (www.springer.com)

Foreword

Remote sensing science and technologies emerged as an important new tool for environmental applications over four decades ago. The environmental applications of remote sensing have gained an irreversible momentum ever since. From early multispectral sensors flown during the NASA Apollo missions in the 1960s that gave birth to the famous LANDSAT series of remote sensing satellite to the present day collection of remote sensing satellites from space agencies across the globe (USA, Russia, Europe, Canada, Japan, India, China, and Brazil, to name a few), the unabated rapid progress in both the sensors and the platforms have provided a major impetus for the use of remote sensing in environmental applications. The results of these applications have been spectacular. Researchers have made groundbreaking observations about the environmental phenomena, and the decision makers have used the information in dialogs over management and policy issues related to the environmental impacts and sustainability.

Over 30 years ago, I stated in one of my early research papers that "*remote sensing has added a new dimension to the analysis and studies of environmental processes, issues and decision-making.*" This statement is still true today as evidenced by the contents of this book *Remote Sensing Applications in Environmental Research edited by Dr. Prashant K. Srivastava, Prof. Saumitra Mukherjee, Dr. Manika Gupta and Dr. Tanvir Islam.* This is indeed a welcome and timely contribution to the growing field of environmental remote sensing.

This book is a compilation of authoritative and comprehensive papers on methods, techniques, and applications as well as illustrative examples of advancements of environmental applications of remote sensing. The editors have organized the book into two parts: Part I is devoted to "classic" applications of remote sensing using various environmental research studies. Part II addresses the advancements in environmental applications of remote sensing. The contributors have covered a wide variety of sensors from a diverse array of remote sensing platforms (Landsat TM and ETM+, MODIS, IRS-LISS, and Hyperspectral) and discussed new approaches to image processing and algorithm development. In addition, a glimpse into the future challenges of environmental applications of remote sensing will offer food for thought for all.

The editors have skillfully presented this detailed and technical content in a user-friendly and coherent format. I am confident that the readers of this book will quickly appreciate the rapid advancements being made in this field. I am positive that the researchers and students alike will gain insights into the new dimensions of applying remote sensing to environmental analysis and understanding. I wish the editors and publishers of this important volume a great success.

Houston, Texas Kamlesh Lulla

Contents

About the Editors

Dr. Prashant K. Srivastava is a Research Scientists at Hydrological Sciences (Code 617), ESSIC/NASA Goddard Space Flight Center, Greenbelt, Maryland, USA. He received his B.Sc. degree in Agriculture Sciences from Institute of Agricultural Sciences, Banaras Hindu University, India in 2004 and M.Sc. degree in Environmental Sciences from School of Environmental Sciences (SES), Jawaharlal Nehru University (JNU), India in 2006. He joined as an Assistant Professor in Department of Environmental Sciences, NVPAS, Sardar Patel University, Gujarat in 2007. In 2010, under the Commonwealth Scholarship and Fellowship Plan (CSFP), he joined Department of Civil Engineering, University of Bristol, Bristol, for his doctoral thesis. His Ph.D. research was focused on the soil moisture retrieval algorithm development from SMOS satellite and mesoscale model for hydrological applications. Following his Ph.D. in 2013, he joined as a research scientist in the ESSIC/NASA Goddard Space Flight Center, Hydrological Sciences Branch on SMAP satellite soil moisture retrieval algorithm development. Dr. Srivastava has been a recipient of many awards including Commonwealth fellowship, UK, CSIR fellowship (twice), MHRD and UGC fellowships from India. He has published more than 48 peer-reviewed journal papers and more than 20 conference papers. Currently, he is serving as an editorial board member of five journals related to remote sensing and environmental sciences. He is currently a member of Indian Society of Geomatics, Indian Society of Remote Sensing, Indian Association of Hydrologists (IAH), International Society for Agrometeorology (INSAM), and International Association of Hydrological Sciences (IAHS).

Dr. Saumitra Mukherjee is Professor of Geology, Remote Sensing and Space sciences at School of Environmental Sciences, Jawaharlal Nehru University, Delhi. He obtained his M.Sc. and Ph.D. in Geology from Banaras Hindu University, Varanasi, India. Prior joining JNU in 1992, Prof. Mukherjee served as Hydrogeologist and Remote Sensing Scientist, Government of India. He has vast experience on satellite like IRS, LANDSAT, SPOT, Indian Remote sensing Satellites, Hyperion, ResourseSat, LIDAR, and SOHO data for Sun–Earth Environment. He has authored seven books in the field of Remote sensing and environmental sciences. He has published more than 80 papers in the peer-reviewed journals with nearly 60 conference contributions. He has supervised 18 doctoral theses and completed 14 collaborative projects funded by National and International agencies, and also involved in the projects funded from NASA on sun–earth connections. Prof. Mukherjee has been designated as UGC Professor of Geology and Remote sensing in view of his contribution in the field of Geology and Remote sensing. He has also served as an executive commonwealth Fellow for University of Liverpool, UK. He is a recipient of many awards including Ministry of Environment and Forest, Government of India on his academic contributions. He is serving as an editorial board member of four journals. Currently, he is a member of number of pioneer organizations in world including American Geophysical Union, European Space Agency, European Geosciences Union, SOC, JPL/CALTECH/NASA, European Fleet for Airborne Research, Canadian Remote sensing Society, Indian Society of Remote Sensing and Executive Fellow (Earth Sciences-India).

Dr. Manika Gupta received her Master degree in Environmental Sciences from School of Environmental Sciences (SES), Jawaharlal Nehru University (JNU), India in 2006. She worked as a research personnel on remote sensing applications from SES, JNU for a year before joining Department of Civil Engineering, Indian Institute of Technology, New Delhi, India for her Ph.D. in 2007 as a UGC fellow. Her Ph.D. focuses on numerical modeling for water resource management. She is also working as a research consultant for satellite based slum area delineation of Jaipur, India. She has collaboration with number of pioneer organizations in the world such as University of Bristol, NOAA, and NASA on remote sensing advancements. Her current research involves assimilation of hyperspectral and microwave (precipitation) data in numerical model for pesticide and soil moisture prediction. Dr. Manika Gupta has been a recipient of many awards including EIPRS (Australia), COSPAR (China), CSIR, ICMR, and UGC fellowships from India. She has published more than 25 peer-reviewed journal papers and more than 15 conference papers. Currently, she is serving as an editorial board member of two journals related to remote sensing and environmental sciences. She is also member of Indian Society of Geomatics, Indian Society of Remote Sensing, Indian Association of Hydrologists (IAH), and International Society for Agrometeorology (INSAM).

Dr. Tanvir Islam received the Ph.D. degree in remote sensing from the University of Bristol, Bristol, UK, in 2012. His Ph.D. research was focused on the remote sensing of precipitation through the use of radar polarimetry, especially towards algorithm developments and data quality improvements. Following his Ph.D., he joined the University of Tokyo as a visiting scientist, more specifically, as part of the NASA/JAXA precipitation measurement missions. Since 2013, he has been with the NOAA Center for Satellite Applications and Research, as a CIRA fellow, working on the development of satellite remote sensing algorithms, with an emphasis on microwave variational inversion techniques. Currently, he is the scientific algorithm developer for the NOAA's Microwave Integrated Retrieval System (MiRS). Dr. Tanvir was the recipient of the Faculty of Engineering Commendation from the University of Bristol for his outstanding Ph.D. thesis and nominated for a University Prize, in 2012, the JAXA visiting fellowship award, in 2012, and the CIRA postdoctoral fellowship award, in 2013. He is member of the American Geophysical Union (AGU), International Association of Hydrological Sciences (IAHS), and American Society of Civil Engineers (ASCE).

Introduction

Prashant K. Srivastava, Saumitra Mukherjee,
Tanvir Islam and Manika Gupta

Remote Sensing Advancements for Environmental Applications

Satellite remote sensing has been transformed in less than 30 years from being a sparse research tool into a commodity product available to a broad user community. However, after three decades of remote sensing advancements, still there is a need for standardized data processing techniques that may take into account the special properties of remote sensing datasets. The major advantage of satellite images is that it varies in spectral, spatial, and temporal resolution, and therefore can be used for variety of applications and provide a more complete view of the observed objects. Hence, seminal views on recent advances in remote sensing techniques from classical to new advancements are very much required.

With increasing numbers of Earth-observing satellites in space, huge volumes of remote sensing data will be produced and there are many more coming in future. Classical remote sensing data processing and distribution methods generally suffer from poor performances and thus make it unsuitable for various end users and real-time forecasting. More detailed analysis and efficient datasets may help the satellite data usage with more confidence. Day-by-day, the earth remote sensing data retrieval, processing, distribution, and application are becoming more challenging and hence require a more detailed analysis on a regular basis. One of the challenges are maintaining the high volumes of satellite datasets and transferring them into a form which is conveniently used by nearly all scientific communities again by utilizing very less space. Custom-made satellite remote sensing data with interactive graphical user interface on a GIS/Web-GIS compatible format is very important for a lot of end users, especially for novice users. On the other hand, more widely distributed Earth-observing remote sensing data in different formats through diversified protocols will result in better usage of future Earth-observing satellite systems. To address these issues, data compressing and preprocessing (sub-setting and sub-sampling), data format conversing (easy accessing data format such as GIS compatible format), GIS and Open GIS applications, and simple real-time data processing for future Earth-observing satellite systems can be a promising approach.

The future of remote sensing lies in making numerous types of accurate, current, and high-resolution remotely sensed data and derived geospatial information products readily available for every area of interest. Scientific community looks forward to the day when datasets are readily available for analysis in nonspecialized software packages. Challenges facing terrestrial and atmospheric remote sensing are unlimited as the different topographies behave differently with particular land surface dynamics and meteorological conditions across the globe at the various spatial and temporal domains. The main goals are to make accurate estimates of selected key sets of geophysical variables, with the intention of either making accurate predictions across time and spatial boundaries of various geographical conditions, or advancing fundamental knowledge through the development of empirical and/or theoretical models. The other challenges are related to our understanding of the geophysical processes, the sensor physics, optical and microwave properties of the surface as well as the sampling capabilities for accurate measurements.

Many new technologies are required for the improvement of individual system capabilities and for the operation of large numbers of sensors. Because of fundamental physical limit, future projection is difficult which therefore creates a question mark on future progress. Therefore, it is required to outline the generalized form of future remote sensing systems and the technologies they will rely on. The abovementioned technological challenges for future global Earth observation can best be met through the combined efforts of the international community. To help make this vision a reality, NOAA/NASA remote sensing experts recommended a number of challenges over the next two decades. According to them, scientific community should focus on improving the accuracy, resolution, timeliness, and ease-of-use of remotely sensed data as well as improving the delivery of remotely sensed data via the Internet, so that users can more easily find and retrieve the exact data and information as per their requirements. NOAA/NASA further stresses on natural and human-induced disaster mitigating strategies by providing high-resolution data of affected areas immediately. There are a number of sensors both on optical and microwave platforms, each with some unique capabilities. If they are to be used synergistically (or fused), it will improve the quantity and quality of information available to the public. Further to software support group, it is suggested to improve the software so that remotely sensed data can be quickly and easily processed and analyzed, and simultaneously yield information needed by professionals.

In this edited volume, the advancements in remote sensing are discussed in terms of the design of techniques able to deal with the high-dimensional nature of the data, and to integrate the spatial and spectral information in a synoptic and more directive way. The recent advancements are discussed in brief along with performances in terms of case studies in different analytical scenarios. All state-of-the-art remote sensing techniques are discussed for specific applications with efficient parallel implementations of some of the algorithms. Synoptically, these parts provide an excellent snapshot and offer a thoughtful perspective on future potentials and emerging challenges in the design of robust remote sensing applications. These advancements are not limited to only communication and extra celestial sciences but now spread over a variety of terrestrial, oceanic, and atmospheric application. In the past few years, there have been many developments in our understanding of optical and microwave remote sensing approaches for environmental applications within land cover, climate change, vegetation, meteorology, hydrology, atmospheric sciences, and oceanic studies. The purpose of this book is to provide a platform for scientists and academicians all over the world to promote, share, and disseminate the new advancements in satellite remote sensing for the abovementioned applications. The chapters are presented in such a way so that they will focus on nearly all issues from classical remote sensing to advanced remote sensing techniques.

Structure of the Book

The book is mainly divided into two parts. The first part covers the classical remote sensing applications while the second part focuses on the advanced remote sensing techniques for the environmental applications. The chapters primarily cover the remote sensing applications in earth and environmental sciences. Furthermore, advancements in line of remote sensing are discussed and provided, such as GIS, GPS, and geospatial techniques.

The classical remote sensing is basically focussed on products or application derived using the previously developed techniques or model, but very important for societal benefits. These techniques form a fundamental for advanced remote sensing development. The classical remote sensing part covers the variety of topics from atmosphere to land applications. The first chapter in this part provides MODIS-based determination of conifer needle flushing (CNF) phenology over boreal-dominant regions. The second chapter deals with Watershed Management using Information System integrating with Remote Sensing and GIS. Over here, attempt has been made to integrate dimensions in Agriculture–Water–Soil–Climate continuum for sustainable management using the WATMIS (Watershed Management Information System) tool. Third chapter discusses land surface temperature (LST) retrieval using spectral index (NDVI), spectral radiance, and surface emissivity using Landsat data for a moist deciduous tropical forest. This study is very useful for forest conservation and management. In fourth chapter, a brief investigation of sensitivity analysis has been performed using SimSphere Land Surface Model, which is very promising for operational products development from Earth Observation (EO) data. The fifth chapter in this part provides geospatial techniques integrated with IRS P6 LISS III data to analyze the precise estimation of the percentage of the soil organic carbon associated with organic matter in soil in Tropical Wildlife Reserve.

The advanced remote sensing part covers the latest techniques developed for environmental applications. The first chapter in this part provides fuzzy unordered rules induction algorithm and j48 decision tree models for spatial prediction of shallow landslides. The landslide conditioning factors derived from remote sensing and GIS are used for the development of landslide hazard maps. Flooding is one of the severe disasters that causes mass demolition of properties and affects human lives. The next chapter attempts to address this problem using the remote sensing and GIS techniques. The submergence zone categorization integrated with Town Planning Scheme (TPS), water level and flood inundation maps are generated on remote sensing and GIS platform. In next chapter, the regional measurements of fire is discussed using the MODIS data. The spatial variations in vegetation fires and carbon monoxide concentrations in South Asia are presented to evaluate its role in environmental pollution. However, in remote sensing, land use/land cover change detection is a very important component. The next chapter provides a brief background of fragstats modelling for driving landscape matrices derived from remotely sensed data (LANDSAT and IRS 1D LISS III) aiming to characterize the historical trends of landscape changes. This study is important for deriving the changes on a long term and will be helpful in mitigating adverse impact of these changes. The tenth technical chapter deals with the hyperspectral data analysis for deriving chlorophyll which is an important component for yield measurements and agricultural crop monitoring for diseases or any stress. In this chapter the Linear Regression Algorithms is used for predicting chlorophyll using the ground spectroradiometer data supplemented with ground verifications. The last chapter covers the artificial intelligence technique integrated with remote sensing for archaeological site management. This chapter opens new paradigm in archaeological site location and managment using high resolution optical satellite based datasets such as from advanced wide field sensor.

Part I
Classical Remote Sensing Applications

Remote Sensing-Based Determination of Conifer Needle Flushing Phenology over Boreal-Dominant Regions

Navdeep S. Sekhon, Quazi K. Hassan and Mohammad M. Kamal

Abstract Coniferous needle flushing [CNF: defined as the date when the tips of at least 75 % fresh buds of white spruce (*Picea glauca*) and/or black spruce (*Picea mariana*) in the surrounding area have reached to a minimum of 2 cm new growth since the start of the growing season] is one of the critical phenological stage in particular to boreal forest. Here, our objective was to evaluate the performance of remotely sensed MODIS-data in determining CNF stage in the Canadian province of Alberta. We employed two predictors primarily using Moderate Resolution Imaging Spectroradiometer (MODIS) data, i.e. (1) accumulated growing degree days (AGDD) and (2) normalized difference water index (NDWI). For determining the thresholds for both of the predictors, we extracted temporal trends AGDD and NDWI during the period of ground-based CNF observations at the lookout tower sites. We found that individual thresholds of AGDD (i.e., 200 degree days) and NDWI (i.e., 0.525) were in better agreements (i.e., ~85 and ~72 % of the times for AGDD and NDWI respectively for ±2 periods of deviation) with the ground-based CNF observation periods. The combination of the two predictors revealed that their logical 'OR' combination produced the overall best agreements (i.e., on an average ~85 % of the times within ±2 periods of deviation).

Keywords Accumulated growing degree days · Normalized difference water index · MODIS data · White spruce (*Picea glauca*) · Black spruce (*Picea mariana*)

N. S. Sekhon · Q. K. Hassan (✉)
Department of Geomatics Engineering, Schulich School of Engineering,
University of Calgary, 2500 University Dr. NW, Calgary, AB T2N 1N4, Canada
e-mail: qhassan@ucalgary.ca

M. M. Kamal
Department of Geography and Planning, University of Saskatchewan,
117 Science Place, Saskatoon, SK S7N 5C8, Canada

P. K. Srivastava et al. (eds.), *Remote Sensing Applications in Environmental Research*,
Society of Earth Scientists Series, DOI: 10.1007/978-3-319-05906-8_1,

1 Introduction

Vegetation phenology is the study of the recurring growth responses of the plants as influenced by seasonal changes in climate. In most of the instances, the climatic variables are primarily responsible for both seasonal and inter-annual variability observed in the phenological events at local, regional, and global scales (Cleland et al. 2007; Dufour and Morin 2010). Here, we are interested in one of the critical boreal phenological stage, i.e., "conifer needle flushing" [CNF: defined as the date when the tips of at least 75 % fresh buds of white spruce (*Picea glauca*) and/or black spruce (*Picea mariana*) in the surrounding area have reached a minimum of 2 cm new growth since the start of the growing season (FFMT 1999)]. The study of CNF is worthwhile because it: (1) defines the transition between spring developmental phases and the peak growing season (Hannerz 1999); (2) indicates the end of proneness to bud damage due to late spring frost (Jones and Cregg 2006; Man et al. 2009); (3) relates to the carbon exchange aspect, as conifers show a transition from source to sink near the completion of the bud developmental stages (Leinonen and Kramer 2002; Richardson et al. 2009); and (4) defines the foliar moisture content and thus has potential in predicting forest fire danger (Wang and Zwiazek 1999; Agee et al. 2002).

In determining the conifer phenological stages, the most commonly employed method is the ground-based observations (e.g., Hannerz 1999; Man and Lu 2010). Over our region of interest, the Alberta Sustainable Resource Development has been acquiring ground-based phenological observations using a network of look-out towers since the 1970. However, these observations may potentially vary from person to person depending on their interpretation skills (Sekhon et al. 2010); and also fail to address spatial variability.

In order to address the spatial dynamics of boreal vegetation phenology, remote sensing-based methods have also been evaluated over various geographic regions (Reed et al. 2009; Tan et al. 2011). These methods mainly rely on detecting the changes in biophysical and/or biochemical characteristics influenced by the climatic variables by exploiting the temporal trends observed in various remote sensing-derived vegetation indices (Xiao et al. 2009; Carrao et al. 2010). These include the use of: (1) normalized difference vegetation index [NDVI: function of reflectance in red and near infrared spectral bands; e.g. (Tan et al. 2011)]; (2) enhanced vegetation index [EVI: function of reflectance in red, blue and near infrared; e.g. (Tan et al. 2011)]; and (3) normalized difference water index [NDWI: function of reflectance in near infrared and short wave infrared spectral bands; e.g. (Delbart et al. 2005)]. The implementations of vegetation indices related to greenness, in particular (i.e., NDVI and EVI) over the boreal forest, indicate that these have large uncertainty in determining the onset of spring due to the presence of snow (Delbart et al. 2005; Sekhon et al. 2010). On the other hand, NDWI (a measure of canopy moisture content) has demonstrated its potential in determining the onset of spring over the boreal forest (Delbart et al. 2005). This is due to the fact that the short wave infrared bands (i.e., used to formulate NDWI) are sensitive to the canopy/plant water content. In general, these vegetation indices are quite capable of determining the phenological stages of spring onset, peak of the growing season, and the end of the growing season (Reed et al. 2009; Xiao et al. 2009). However, the application of any of the remote sensing-derived indices in determining the phenological stage of our interest (i.e., CNF) has not been reported in the literature so far.

The above described remote sensing-based methods rely on the 'response' change caused by the occurrence of a phenological stage. Alternatively, the phenological stages can also be determined from the cause of its occurrence (e.g., the influence of favourable temperature regimes). For example: Hassan and Rahman (2013a, b) employed accumulated growing degree days (AGDD) derived primarily from Moderate Resolution Imaging Spectroradiometer (MODIS)-based surface temperature (T_S) in determining both deciduous and understory grass greening phenology over boreal-dominant forest. Thus, we propose to delineate CNF stage using two variables, such as AGDD and NDWI over the boreal-dominant forest in the Canadian province of Alberta. The specific objectives are to: (1) determine thresholds of cause (i.e. AGDD) and response (i.e. NDWI) for predicting CNF during 2006 using ground-based CNF observations; (2) compare CNF predictions (i.e. based on the determined thresholds for AGDD, NDWI, and both combined together logically) with the ground-based observations for the period 2007–2008; and (3) generate CNF occurrence map using best predictor determined in objective (2), and discuss the spatial dynamics.

2 Materials and Methods

2.1 Description of the Study Area and Data Requirements

We considered the conifer-dominant forested regions in the Canadian province of Alberta as our study area (see Fig. 1 for more details). Geographically, Alberta falls in between parallels of 49–60°N latitudes, and 110–120°W longitudes. It experiences a continental climatic regime with cold winters and relatively short and moderately warm summers with the following characteristics (Dowing and Pettapiece 2006): (1) average annual temperatures vary in the range −3.6 to +4.4 °C, (2) summer average temperatures vary in the range 8.7–18.5 °C, (3) winter average temperatures vary in the range −25.1 to −9.6 °C, and (4) average annual precipitation is in between 333 and 989 mm. On the basis of climatic regimes, soil type, topography, and vegetation, Alberta is being classified into 21 natural subregions (Dowing and Pettapiece 2006). Among these, 9 natural subregions located in the northern and south-western regions (marked by * sign in Fig. 1b) are conifer-dominant ones.

Here, we employed several MODIS-derived datasets, such as: (1) 8-day composites of AGDD maps at 1 km resolution during the period 2006–2008 available

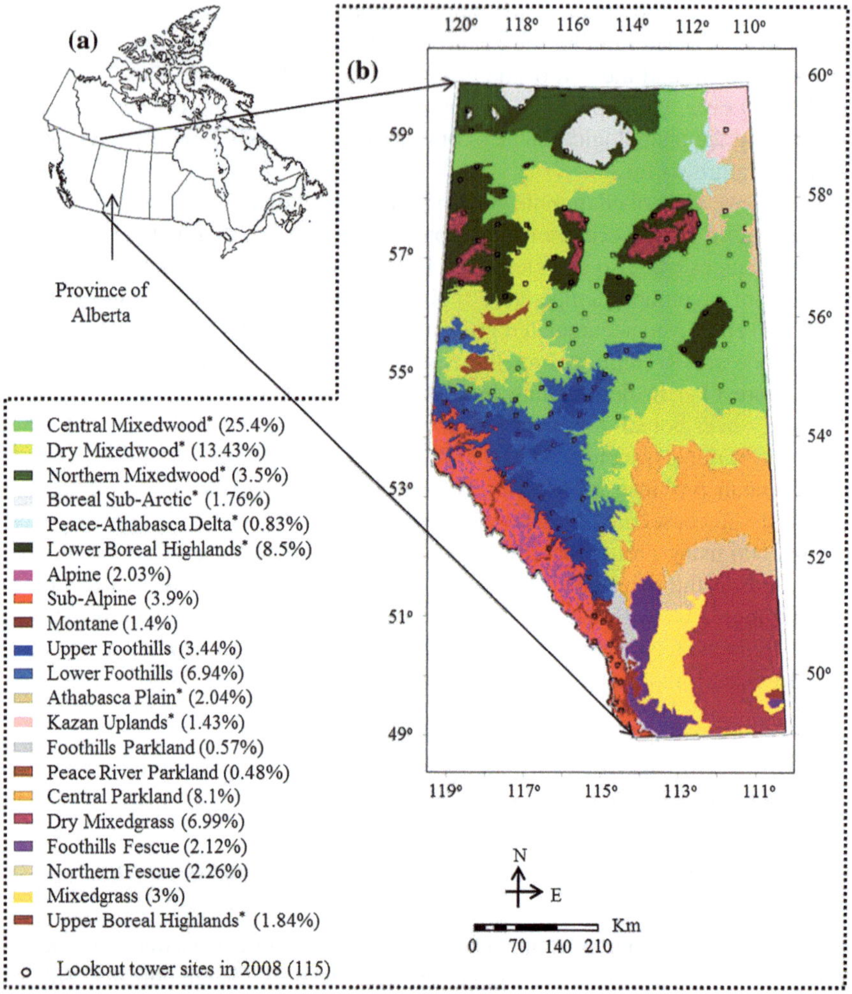

Fig. 1 Geographical extent of the study area in Canadian extent (**a**); and its constituting 21 natural subregions along with the locations of the lookout tower sites of conifer needle flushing observation during 2008 shown in *circles* (**b**). The *asterisk* marked subregions are conifer-dominant ones

from Hassan and Rahman (2013a); (2) 8-day composites of NDWI maps at 500 m resolution during the period 2006–2008 available from Sekhon et al. (2010), which were calculated as a function of surface reflectance from near infrared and short wave infrared spectral band centred at 2.13 μm; and (3) annual composite of land cover (i.e., MCD12Q1 v.004) at 500 m resolution during 2007. In addition, we also employed ground-based CNF observations for the species of white and/or black spruce at ~118 lookout tower sites operated by Alberta Sustainable Resources Development Alberta (see Fig. 1 for location information) for the period 2006–2008. At each of the sites, approximately 20–100 trees were sampled to declare the CNF stage (Dylan Heerema: a veteran lookout tower operator, Personal Communication). As, these ground-based CNF data were recorded in the form of day of year (DOY = 1 to 365 or 366), it required conversion to the equivalent period of MODIS data (i.e. 8-day imaging epochs) for comparison with MODIS-derived CNF stages. For this purpose, we employed the following expression as described in Sekhon et al. (2010):

$$P = \left[\left(\frac{DOY - 1}{8} \right) + 1 \right] \tag{1}$$

where, P (=1–46) is the corresponding equivalent imaging period of MODIS. The value of P is always integer even if the calculated values is a floating number, e.g., P = 15 if Eq. 1 produces values in between 15 and 15.875.

2.2 Generation of AGDD Maps

We obtained 8-day composites of GDD maps at 1 km resolution from an earlier study (Hassan and Rahman 2013a). In this process, the required inputs were: (1) MODIS-based 8-day composites of T_S images at 1 km resolution (i.e., MOD11A2 v.005), and (2) 8-day mean air temperature ($\overline{T}_a$) at 182 weather stations across the study area calculated from their respective daily-values. The steps of generating GDD/ADDD could be summarized as follows. Firstly, the 8-day composites of MODIS-based instantaneous T_S-values were converted into $\overline{T}_a$-values by use of an empirical model (i.e., $\overline{T}_a = 0.61 * T_S + 103.66$, where the temperatures were in Kelvin); and then calculated GDD-values using the following expression:

$$GDD = \begin{pmatrix} \overline{T}_a - T_{base} & \text{if} \quad \overline{T}_a > T_{base} \\ 0 & \text{otherwise} \end{pmatrix} \tag{2}$$

where, T_{base} is base temperature with a magnitude of 5 °C (=278.15 K), as the trees may require this minimum air temperature to initiate growing conditions (Hassan and Bourque 2009; Rossi et al. 2007). Secondly, these 8-day composites of GDD maps were added at each of the 8-day periods and progressed throughout the growing season in order to produce the "AGDD" maps.

2.3 Determination of AGDD and NDWI Thresholds for CNG Occurrence

As a first step, we extracted temporal trends of both AGDD and NDWI during the years 2006–2008 at the lookout tower sites where the ground-based CNF records were available. These datasets were then divided into two groups: (1) calibration dataset consisting of ~34 % data points (i.e., all data from 2006); and (2) validation dataset consisting of ~66 % data points (i.e. all data from the years 2007–2008). During the calibration phase, we performed the following steps:

1. Extracted both of the AGDD and NDWI-values at each 8-day intervals at each of the lookout tower sites. During every 8-day period, we then calculated the average and standard deviation-values for both AGDD and NDWI by considering their respective values at all of the lookout tower sites.
2. Considered the average values associated with AGDD and NDWI during the CNF ground-based observation period as the predictor-specific initial "threshold-values" for CNF determination.
3. Evaluated the sensitivity of the initial threshold-values for both of the AGDD and NDWI-values in determining the CNF stage upon varying in the bound "initial threshold ±1 standard deviation" as most of the data were found to be in this range. Within the bound, we evaluated the thresholds at each "1/3 standard deviation" intervals. The best thresholds were then identified when the deviations between the predicted and observed CNF period were the least. Note that − and + deviations mean the early and delayed predictions respectively in relation to the ground-based CNF observation periods throughout the chapter.

In validation phase, we predicted the CNF period during the years 2007–2008 using the best thresholds for AGDD and NDWI obtained in the calibration phase. Then we calculated the deviations between the predicted and observed CNF periods to determine the suitability of the proposed method.

2.4 Integration of both AGDD and NDWI Threshold for CNG Occurrence

We evaluated how the integration of the causal (i.e., AGDD) and response (i.e., NDWI)-based predictors could enhance the determination of the CNF stage. In the process of integrating both AGDD and NDWI-values in determining the CNF stage, we considered the best thresholds of AGDD and NDWI determined in Sect. 2.3. The CNF period was predicted in two ways: (1) if at least one of the predictors reached to the corresponding threshold-value (i.e., logical 'OR' condition between the predictors); and (2) both of the predictors would reach to the threshold-values (i.e., logical 'AND' condition between the predictors). Upon implementing both methods, we then compared the predicted CNF period with the ground-based CNF observations by analysing the deviations between them.

2.5 Mapping of CNF Using the Best Prediction Criteria

Among all of the evaluated predictors (i.e., AGDD and NDWI individually as described in Sect. 2.3; and the integration of AGDD and NDWI as described in Sect. 2.4), we implemented the best predictor in mapping CNF stage over the conifer-dominant natural subregions (i.e., marked with * symbols beside the name of the natural subregions in Fig. 1b).

3 Results and Discussion

3.1 Determination of AGDD Threshold for CNF Occurrence

Figure 2a shows averaged temporal trends of AGDD in 2006 upon considering all of the lookout tower sites (i.e., 114 in total). The initial AGDD threshold (i.e., average at all of the lookout tower sites), was found to be 225 degree days with a standard deviation of ±75 degree days during the period of ground-based CNF observations (averaged ground-based CNF observation period shown in Fig. 2a with a dotted vertical line). The variability observed in the AGDD thresholds might be associated with one or assimilation of the following factors: (1) water stress (Royce and Barbour 2001); (2) climatic conditions of previous season (Gamache and Payette 2004; Dufour and Morin 2010); and (3) genetic differences among inter and intra-species (Weilgolaski 1999; Tanja et al. 2003). The frequency distribution of the AGDD thresholds revealed that they were in between 150 and 300 degree days (i.e., initial threshold ±1 standard deviation) for approximately 71 % of the times (see Fig. 2b). Upon varying the threshold in the range of "initial threshold ±1 standard deviation" at each 1/3 standard deviation intervals, we found that the threshold of 200 degree days were able to produce the best overall agreements at 0 (i.e., 30.7 %), ±1 (i.e., 68.4 %), and ±2 (i.e., 85.9 %) period of deviation (see Fig. 2c for more details). The determined AGDD threshold of 200 degree days in this study was consistent with other studies. For examples: (1) the mean AGDD-values of '*bud flushing*' (i.e., the appearance of the buds before the CNF stage) for white and black spruce were on an average 91 and 150 degree days (with 5 °C base temperature) respectively in northern Ontario (O'Reilly and Parker 1982); (2) AGDD requirements for needle flushing for various clones of Norway spruce (*Picea abies*) in Sweden were 150–250 degree days (Hannerz 1999); (3) AGDD (with base temperature 5 °C) requirements were on an average to be 150 degree days with standard deviation of 56 degree days (with a median of 179 degree days) for white spruce under various controlled temperature treatments (Man and Lu 2010). Note that all these studies were conducted under experimental setup; while our study represented the observations over naturally occurring stands.

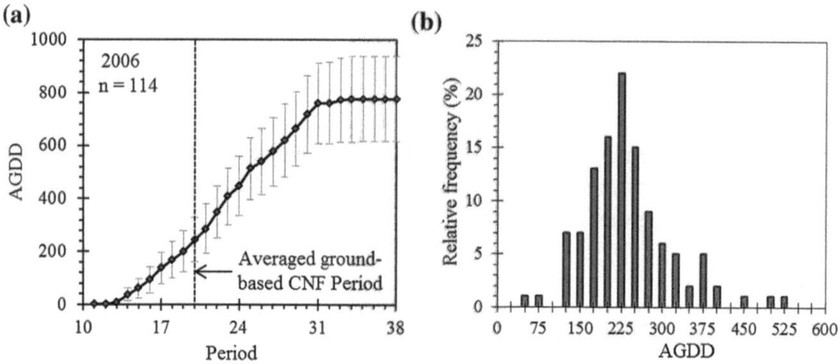

(c)

% out of 114 lookout towers (2006)

AGDD Threshold	Deviation (in periods) 0	± 1	± 2	± 3	± 4	± 5	≥ ± 6
150	14	32.5	50	69.3	93.9	100	100
175	21	66.6	84.1	92.9	98.2	100	100
200	30.7	68.4	85.9	93	100	100	100
225	18.4	58.8	84.2	93	100	100	100
250	14	46.4	73.6	93.8	98.2	100	100
275	10.5	31.5	63.1	90.3	97.3	98.2	100
300	2.6	2.6	10.5	28.1	37.7	50	100

Fig. 2 Determination of AGDD threshold for CNF occurrence: **a** averaged temporal trends of AGDD at all of the lookout tower sites during 2006, and the *dotted vertical line* shows the averaged ground-based CNF observation period; **b** relative frequency of the AGDD at all of lookout tower sites; and **c** implementation of various AGDD threshold to determine the best threshold in predicting the CNF period

Table 1 Relation between AGDD-based prediction and ground-based CNF observation period at the lookout tower sites during 2007–2008 using the best AGDD threshold of 200 degree days

Year	No. of lookout tower sites	Deviation (in periods) 0	±1	±2	±3	±4	±5	±6	≥±7
2007	118	21.2	61	89.8	98.3	100	100	100	100
2008	115	19.1	53	80.8	94.7	98.3	98.3	100	100
2007–2008	116	20.2	57.1	85.4	96.6	99.2	99.2	100	100

The implementation of the final AGDD threshold (i.e., 200 degree days) during the period 2007–2008 revealed that the deviations were within ±2 period for reasonable amount of the times (i.e., 89.8 % in 2007, 80.8 % in 2008, and 85.4 % in 2007–2008) (see Table 1 for more details). The consideration of ±2 periods deviation in assessing the quality of the vegetation phenological stage of interest (e.g., SGN stage) was quite acceptable as applied in other studies (Sekhon et al. 2010). The larger deviations (i.e., >±2 periods) might have happened due to the following reasons: (1) the ground-based observations were subjective, and highly dependable on the interpretation of the observer (Sekhon et al. 2010); (2) spatial resolution of MODIS-based prediction might not be always at the same scale of ground observation for all of the cases (Fisher and Mustard 2007); and (3) the implementation of a global AGDD threshold over the entire study area might be unable to capture all of the variability (Man and Lu 2010; O'Reilly and Parker 1982).

3.2 Determination of NDWI Thresholds for CNF Occurrence

Figure 3a shows an average temporal trend of NDWI-values by extracting their values at all of the lookout tower sites (i.e., 114 in total) during 2006. We observed that the NDWI-values were relatively stable (i.e., maintaining an average of ~0.7) during the winter season (i.e., between 1–11 period; 1 Jan.–29 March) due to the prevailing snow covers; and followed by a decreasing trend as a result of snow melting and reached to a minimum value [i.e., defined as the end of snow melting (Delbart et al. 2005) or SGN stage (Sekhon et al. 2010)] before it started to increase again. The average of NDWI-values in comparison with the average ground-based CNF observation periods at 114 lookout tower sites (shown in Fig. 3a using dotted vertical line) during 2006 was 0.523 with a standard deviation of 0.165. The observed variability in the initial NDWI threshold might also be associated with the similar relevant factors described in Sect. 3.1. The frequency distribution of the NDWI thresholds revealed that they were in between 0.375 and 0.675 (i.e., approximately initial threshold ±1 standard deviation) for approximately 60 % of the times (see Fig. 3b). Upon varying the threshold in the range of "initial threshold ±1 standard deviation" at each 1/3 standard deviation intervals, we observed that the NDWI threshold values of 0.525 and 0.575 were predicting the CNF with similar agreements (i.e., ~54 % at ±1 period deviation; see Fig. 3c). We considered 0.525 as the final NDWI threshold which was the closet to the initial threshold of 0.523 and was best at ±2 period deviation level (i.e., ~70 % of times). Note that it was not possible to compare the determined NDWI threshold value with any other reported studies, as the implementation of NDWI temporal dynamics in determining the CNF stages were so far not found in the literature.

The implementation of the final NDWI threshold (i.e., 0.525) during the period 2007–2008 revealed that the deviations were within ±2 period for reasonable amount of the times (i.e., 67.8 % in 2007, 75.6 % in 2008, and 71.7 % in 2007–2008) (see Table 2 for more details). The rationales of observing relatively large discrepancies (i.e., ≥±3 periods deviation) between ground-based observation and predicted CNF stage would follow the same explanations discussed in Sect. 3.1.

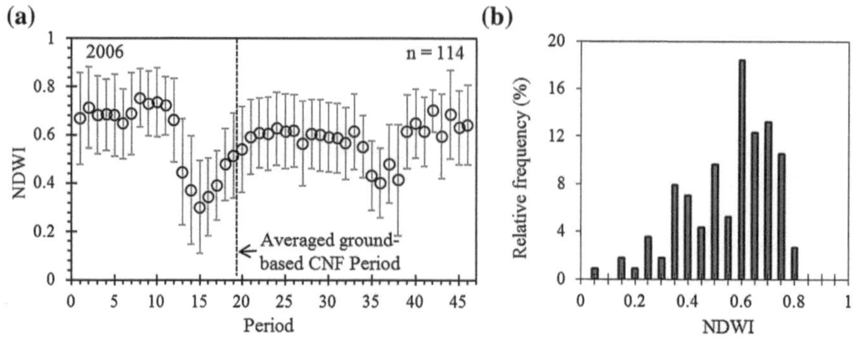

(c)

% out of 114 lookout towers (2006)

NDWI Threshold	Deviation (in periods)						
	0	± 1	± 2	± 3	± 4	± 5	≥ ± 6
0.375	8.8	36	63.2	77.2	87.7	94.7	100
0.425	9.6	42.1	64.9	78.9	86.8	93.9	100
0.475	15.8	50	69.3	81.6	86	92.1	100
0.525	21.1	54.4	70.2	80.7	86	88.6	100
0.575	17.5	54.4	64.9	73.8	78.9	84.2	100
0.625	12.3	40.4	53.5	65.8	69.3	71.9	100
0.675	12.3	28.1	43.9	50.9	56.1	62.3	100

Fig. 3 Determination of NDWI threshold for CNF occurrence: **a** averaged temporal trends of NDWI at all of the lookout tower sites during 2006, and the *dotted vertical line* shows the averaged ground-based CNF observation period; **b** relative frequency of the NDWI at all of lookout tower sites; and **c** implementation of various NDWI threshold to determine the best threshold in predicting the CNF period

Table 2 Relation between NDWI-based prediction and ground-based CNF observation period at the lookout tower sites during 2007–2008 using the best NDWI threshold of 0.525

Year	No. of lookout tower sites	Deviation (in periods)							
		0	±1	±2	±3	±4	±5	±6	≥±7
2007	118	25.4	57.6	67.8	78.8	82.2	87.3	90.7	100
2008	115	33.9	65.2	75.6	83.4	85.1	88.6	92.1	100
2007–2008	116	29.6	61.4	71.7	81.1	83.7	88	91.4	100

3.3 Integration of both AGDD and NDWI Thresholds

The results from combining both of the predictors using logical ‘OR’ and ‘AND’ conditions in predicting CNF stage are shown in Table 3. It revealed that the logical ‘OR’ produced the best results at ±2 period deviation (i.e., ~85 % of the times). In fact, this combination enhanced the predictability at: (1) 0 period deviation-level (i.e., 28.8 % of the times) in comparison to AGDD threshold (i.e., 20.2 %); (2) ±1

Table 3 Relation between the combination of AGDD- and NDWI-based prediction and ground-based CNF observation period at the lookout tower sites during 2007–2008

Year	No. of lookout tower sites	Criteria	Deviation (in periods)							
			0	±1	±2	±3	±4	±5	±6	≥±7
2007	118	'OR'	22.9	63.6	83.1	94.1	96.7	97.5	98.3	100
		'AND'	23.7	55.1	74.6	83.1	85.7	89.8	92.5	100
2008	115	'OR'	34.8	73.9	87.8	93.9	96.5	96.5	98.2	100
		'AND'	18.3	44.4	68.8	84.4	87	90.5	94	100
2007–2008	116	'OR'	28.8	68.7	85.4	94	96.6	97	98.3	100
		'AND'	21	49.7	71.6	83.6	86.2	90.1	93.1	100

period deviation-level (i.e., 68.7 % of the time) in comparison to both AGDD (i.e., 57.1 % of the time) and NDWI thresholds (i.e., 61.4 % of the time). On the contrary, the logical 'AND' between AGDD and NDWI were 21, 49.7, and 71.6 % of the times at 0, ±1, and ±2 period deviation respectively. The possible reason of the better results in logical 'OR' might be associated with the shifting of positive deviations in AGDD-based predictions towards the zero deviation as a result of early prediction by NDWI threshold. It would be the case where the conifer stands might respond to relatively lower AGDD for needle flushing.

3.4 Spatial Dynamics of CNF Across the Landscape

As we found the best agreements by use of logical 'OR' between the predictors, we employed it in generating a CNF map over the conifer-dominant pixels during 2008 shown in Fig. 4. Note that the model predictions are specific to conifer species of white and/or black spruce; while the map shows predictions for the entire conifer stands, assuming these two species are to be the dominating ones across the landscape. We observed that the CNF stage was taken place in between 153 and 176 DOY (i.e., 1–24 June in 2008) for ~75 % of times (see Fig. 4). The spatial dynamics of CNF were summarized as follows:

- The relatively earlier CNF occurrences (≤152 DOY, i.e., before 01 June 2008) were observed mainly in the natural subregions of dry mixedwood, lower foothills, and portion of central mixedwood in the south. These are related to the fact that these regions experience relatively warm temperature regimes (Dowing and Pettapiece 2006).
- On the other hand, the relatively delayed CNF occurrences (≥177 DOY, i.e., after 24 June in 2008) were found to be in the high elevation regions (i.e., >750 m above the mean sea level), such as, Rocky Mountains, Caribou mountain, and Birch mountain (see Fig. 4). These delayed CNF occurrences are expected as high elevations experience relatively cooler temperature regimes and longer snow cover periods (Hassan et al. 2007; Sekhon et al. 2010).

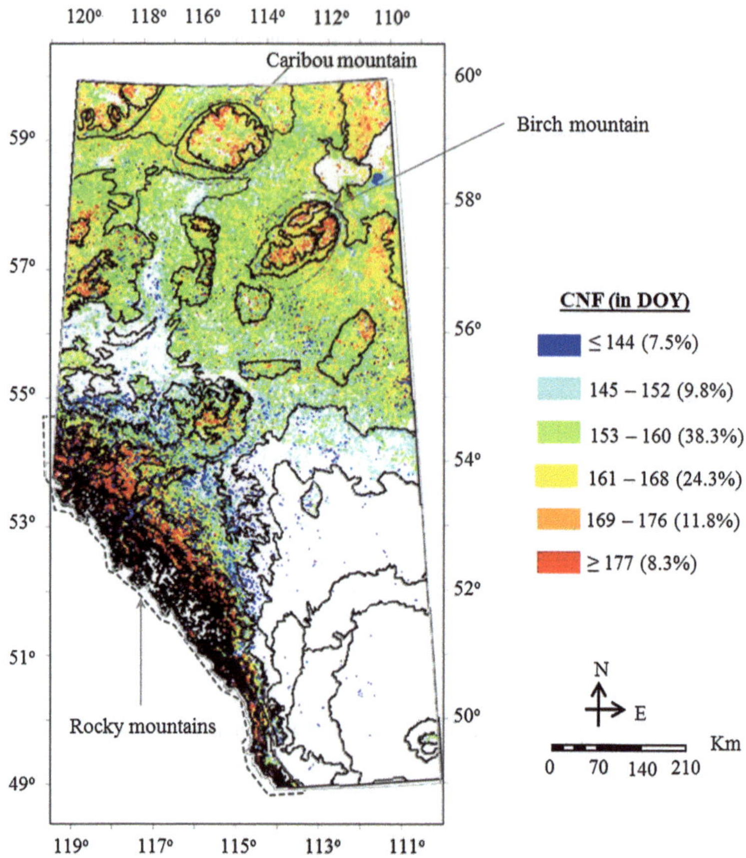

Fig. 4 Spatial dynamics of CNF during 2008 over conifer-dominant 9 natural subregions using the best predictor (i.e., logical 'OR' between AGDD and NDWI thresholds), where the *white* area represents non conifer-dominant natural subregions in Alberta

- In general, the timing of the CNF occurrences was found to be increasing towards north direction. It is consistent with decreasing tendency of the temperature and snow-melt regimes in the northern hemisphere in the same direction (Hassan et al. 2007; Sekhon et al. 2010; Akther and Hassan 2011).

4 Concluding Remarks

In this study, we demonstrated the potential of mapping CNF stage using MODIS-based two predictors (i.e., AGDD and NDWI) over boreal-dominant regions in the Canadian province of Alberta. In the process, we determined study-area specific thresholds for both AGDD and NDWI-values individually for CNF predictions.

We also integrated the two predicting thresholds using logical ‘OR’ and ‘AND’ combinations. Among all of the methods, we found that the logical ‘OR’ produced the best agreements (i.e. $\sim$85 % of times within $\pm$2 period of deviation). The application of the methods as described here establishes the potential of such techniques in delineating the phenology over the boreal regions. However, these methods may require prior investigation and further validation before implementing over other biomes.

Acknowledgements This study was partially supported by an NSERC Discovery Grant provided to Dr. Hassan. The authors would like to acknowledge: (1) NASA for providing the MODIS data; and (2) Alberta Department of Sustainable Resource Development for providing ground-based CNF observation data.

References

Agee JK, Wright CS, Williamson N et al (2002) Foliar moisture content of pacific northwest vegetation and its relation to wildland fire behaviour. Forest Ecol Manage 167:57–66

Akther MS, Hassan QK (2011) Remote sensing based estimates of surface wetness conditions and growing degree days over northern Alberta, Canada. Boreal Environ Res 16:407–416

Carrao H, Goncalves P, Caetano M (2010) A nonlinear harmonic model for fitting satellite image time series: analysis and prediction of land cover dynamics. IEEE Trans Geosci Remote Sens 48:1919–1930

Cleland EE, Chiune I, Menzel A et al (2007) Shifting plant phenology in response to global change. Trends Ecol Evol 22:357–365

Delbart N, Kergoats L, Toan TL et al (2005) Determination of phenological dates in boreal regions using normalized difference water index. Remote Sens Environ 97:26–38

Dowing DJ, Pettapiece WW (eds) (2006) Natural regions and subregions of Alberta. Natural Regions Committee, Government of Alberta, Alberta, Canada. Publication No. T/852

Dufour B, Morin H (2010) Tracheid production phenology of *Picea mariana* and its relationship with climatic fluctuations and bud development using multivariate analysis. Tree Physiol 30:853–865

FFMT (Forest Fire Management Terms) (1999) Forest Protection Division: Alberta Land and Forest Service. http://www.srd.alberta.ca/Wildfire/WildfireOperations/documents/ForestFireManagementTerms-Glossary-1999.pdf. Accessed 20 Nov 2012

Fisher JI, Mustard JF (2007) Cross-scalar satellite phenology from ground, Landsat, and MODIS data. Remote Sens Environ 109:261–273

Gamache I, Payette S (2004) Height growth response of tree line black spruce to recent climate warming across the forest-tundra of eastern Canada. J Ecol 92:835–845

Hannerz M (1999) Evaluation of temperature models for predicting bud burst in Norway spruce. Can J Forest Res 29:9–19

Hassan QK, Bourque CP-A, Meng F-R et al (2007) Spatial mapping of growing degree days: an application of MODIS-based surface temperatures and enhanced vegetation index. J Appl Remote Sens 1:013511:1–013511:12

Hassan QK, Bourque CP-A (2009) Potential species distribution of balsam fir based on the integration of biophysical variables derived with remote sensing and process-based methods. Remote Sens 1:393–407

Hassan QK, Rahman KM (2013a) Applicability of remote sensing-based surface temperature regimes in determining deciduous phenology over boreal forest. J Plant Ecol 6:84–91

Hassan QK, Rahman KM (2013b) Remote sensing-based determination of understory grass greening over boreal forest. J Appl Remote Sens 7:073578:1–073578:10
Jones GE, Cregg BM (2006) Screening exotic firs for the Midwestern United States: interspecific variation in adaptive traits. HortScience 41:323–328
Leinonen I, Kramer K (2002) Applications of phenological models to predict the future carbon sequestration potential of boreal forests. Clim Change 5:99–113
Man R, Kayahara GJ, Dang QL, Rice JA (2009) A case of severe frost damage prior to budbreak in young conifers in Northeastern Ontario: consequence of climate change? Forest Chron 85:453–462
Man R, Lu P (2010) Effects of thermal model and base temperature on estimates of thermal time to bud break in white spruce seedlings. Can J Forest Res 40:1815–1820
O'Reilly C, Parker WH (1982) Vegetative phenology in a clonal seed orchard of *Picea glauca* and *Picea mariana* in northwestern Ontario. Can J Forest Res 12:408–413
Reed BC, Schwartz MD, Xiao X (2009) Remote sensing phenology: status and the way forward. In: Noormets A (ed) Phenology of ecosystem processes. Springer Science+Business Media, New York, pp 231–246
Richardson AD, Hollinger DY, Dail DB et al (2009) Influence of spring phenology on seasonal and annual carbon balance in two contrasting New England forests. Tree Physiol 29:321–331
Royce EB, Barbour MG (2001) Mediterranean climate effects. II. Conifer growth phenology across a Sierra Nevada ecotone. Am J Bot 88:919–932
Rossi S, Deslauriers A, Anfodillo T et al (2007) Evidence of threshold temperature for xylogenesis in conifers at high altitudes. Oecologia 152:1–12
Sekhon NS, Hassan QK, Sleep RW (2010) Evaluating potential of MODIS-based indices in determining "Snow Gone" stage over forest-dominant regions. Remote Sens 2:1348–1363
Tan B, Morisette JT, Wolfe RE et al (2011) An enhanced TIMESAT algorithm for estimating vegetation phenology metrics from MODIS data. IEEE J Sel Topics Appl Earth Observ Remote Sens 4:361–371
Tanja S, Berninger F, Vesala T et al (2003) Air temperature triggers the recovery of evergreen boreal forest photosynthesis in spring. Glob Change Biol 9:1410–1426
Wang Y, Zwiazek JJ (1999) Spring changes in water relations, gas exchange, and carbohydrates of white spruce (*Picea glauca*) seedlings. Can J Forest Res 29:332–338
Weilgolaski F-E (1999) Starting dates and basic temperatures in phenological observations of plants. Int J Biometeorol 42:158–168
Xiao X, Zhang J, Yan H et al (2009) Land surface phenology: convergence of satellite and CO_2 eddy flux observations. In: Noormets A (ed) Phenology of ecosystem processes. Springer Science+Business Media, New York, pp 247–270

Information System for Integrated Watershed Management Using Remote Sensing and GIS

P. D. Aher, J. Adinarayana, S. D. Gorantiwar and S. A. Sawant

Abstract Watershed management is an endowed approach to mitigate the gap between demand and supply of water and other natural resources, particularly in the fragile arid and semi-arid tropics (SAT). As this is a complex phenomenon, there is need for a reliable Information System/Decision Support System (DSS). Watershed Management Information System (WATMIS) is a viable and generic toolkit for integrated watershed planning and management of its natural resources using multiple technologies like Geographical Information System (GIS), Remote Sensing (RS), Global Positioning System (GPS), hydrological modelling and soft computing tools. In this system, an attempt has been made to integrate dimensions in Agriculture–Water–Soil–Climate continuum for sustainable management of land and water resources judiciously. The application of WATMIS will be useful to various stakeholders such as agriculturists, rural extension community and water resources managers for better decision making.

Keywords Decision support system · Geographical information systems · Global positioning systems · Hydrological modelling · Remote sensing · Soft computing tools · Watershed management and planning

P. D. Aher (✉) · J. Adinarayana · S. A. Sawant
Centre of Studies in Resources Engineering, Indian Institute of Technology Bombay, Powai 400076, Mumbai, India
e-mail: prajakta_aher@iitb.ac.in

S. D. Gorantiwar
Department of Irrigation and Drainage Engineering, Mahatma Phule Agricultural University, Rahuri 413722, India

P. K. Srivastava et al. (eds.), *Remote Sensing Applications in Environmental Research*, Society of Earth Scientists Series, DOI: 10.1007/978-3-319-05906-8_2,

1 Introduction

Inconsistencies and competition over shared water resources between various sectors such as agriculture, industry and domestic sector make it vital component of arid and semi-arid tropics (SAT). Water is becoming scarcer and every fore-warning alert indicates that it will become even more critical in the future. With socio-economic development, contradiction becomes conspicuous between necessitate for water and its limited resources, and it is a matter of concern to the watershed community, such as water resource researchers, scientists and policy makers. Therefore, sustainable water management is a crucial need of the hour. Hence, it becomes important to apply the emerging tools and technologies for ubiquitous watershed management.

1.1 Why Management of Natural Resources on Watershed Basis?

Soil, water and vegetation are the most vital natural resources for sustainable development and management, and hence should be handled and managed effectively, collectively and simultaneously. Managing the natural resource with sustainable approach is a rational phenomenon in its natural region. In this approach, the natural regions are invented to be in terms of the flow of water, which influences almost all fields of the environment, where the regions are diversified as basin, catchment, sub-catchment, macro watershed (>50,000 ha), sub-watershed (10,000–50,000 ha), milli-watershed (1,000–10,000 ha), micro watershed (100–1,000 ha), mini watershed (1–100 ha) (Nair 2009). However, a particular extent/size of a region is imperative with regard to the aim of its development. Size will also be affected by the possible major components of a development such as afforestation, cultivation practices, etc. Keeping in view the local conditions and completion of the project within a reasonably short time, an average size of 2,000 ha is considered rational for agricultural development with regard to ease of surveys and investigations and effective planning. In the present research work, a watershed has been taken as the smallest planning unit, as it conveniently and efficiently represents continuum of three vital natural resources i.e. soil, water and vegetation.

Watershed management programme has emerged as a sustainable strategy to conserve the natural resources i.e. water, forest and soil in an integrated manner particularly in the rainfed and drought areas (Roy 2005). Planning and management of natural resources at micro level of the watershed where there is a high spatio-temporal variability in Gio-physical and socio-economic variables, particularly in the fragile arid and semi-arid tropics (SATs), is the crucial need of the hour (Aher et al. 2012). The real challenge on water resources planning at a micro level is to assess the quantum of water demand and availability caused due to unavailability of adequate database. Watershed based planning through augmentation of modern techniques such as remote sensing (RS) and Geographic Information System (GIS), for modelling the availability of water resources and sectoral demand is being considered as the most appropriate approach.

1.2 Role of Geographic Information System (GIS) and Remote Sensing (RS) in Watershed Management

A Geographical Information System (GIS) can be defined as a system, which facilitates the storage and intelligent use of geographic data and human activities (Srivastava 2003). The essential features of GIS are the use of sophisticated computer hardware and software to collect, store, manipulate and process for geographic data (Singh 2010). GIS is a tool that allow for the processing of spatial data into information (Samarakoon 2005). GIS has the ability to manipulate spatial data and corresponding attribute information to integrate different types of data in a single analysis at high speed, which is unmatched with manual methods (Rashed et al. 2006).

GIS provides a digital representation of landform which could be used in hydrological modelling. The database available in GIS environment facilitates assimilation of different thematic datasets to understand interrelationships. GIS plays an important role in information management, analysis, and providing solutions to the planning of natural resources. The application of GIS for land use surveys and mapping is gaining importance, largely because of its ability to provide rapid and reliable data within a given time framework (Jain 1996). Many GIS-based watershed applications have been developed since the early 1990s due to advances in desktop GIS capabilities, programming languages, and data availability (Strager et al. 2010). In addition, it facilitates integration of socio-economic information with the resources data to understand the local needs. Once the potential of resources and development needs of the watershed are understood, it is possible to evolve specific action plan for development of land and water resources. It plays a key role in implementing soil and water conservation practices that are essential for sustainable agriculture production.

GIS offers technologically suitable method for land resource assessment, delineating different land use patterns, flood management, irrigation water management, and assessment and monitoring of environmental impact of watershed projects. It is also useful in delineating hydro-morphological units in the area to decide suitable sites for land and water harvesting structures in the problematic sites.

Remote sensing is the non-contact recording of information from various electro-magnetic spectrum regions by means of instruments such as cameras, scanners, lasers, linear arrays and/or area arrays located on the ground or arial platforms (Jensen 2007) and the analysis of the acquired information by means of visual and digital image processing (Sabins 1987). Remote sensing, with or without GIS technology, has emerged as an indispensable scientific tool for mapping and planning of natural resources (Vittala et al. 2008; Mahajan and Panwar 2005; Bryan et al. 2011; Burkhard et al. 2012). It plays a hastily escalating role in the field of hydrology and sustainable water resources development and management. These techniques have been extensively applicable in nearly all fields of watershed aspects, like, estimation of evapotranspiration (Bashir et al. 2008; Elhag et al. 2011), soil erosion (Vemu and Pinnamaneni 2011; Esteves et al. 2012; Conoscenti et al. 2013), rainfall runoff modelling (Shrivastava et al. 2004; Rawat et al. 2011; Kim et al. 2012; López-Vicente et al. 2013), flood management (Mason et al. 2003; Park and Hur 2012; Steinfeld et al. 2013) and irrigation water management (Saidi et al. 2009; Georgoussis et al. 2009; Nahry et al. 2011; Liyantonoa et al. 2013).

1.3 Decision Support System in Watershed Management

Decision Support System (DSS) are a model-based set of procedures for processing data and judgments to assist a manager in his/her decision (Little 1970). Adelman (1992) has defined DSS as interactive computer programs that utilize analytical methods, such as decision analysis, optimization algorithms, program scheduling routines, and so on, for developing models to help decision makers to formulate alternatives, analyze their impacts, and interpret and select appropriate options for implementation. DSS is a computer based system of integration of database, models and user interface which are programmed for easily interpretable results to aid the decision makers (Walsh 1993). A DSS can be designed on the basis of user application need as an individual stand-alone or information service based. A stand-alone DSS can be run on a computer dedicated to DSS task where computer acts as desktop microcomputer/high performance workstations or on a multiple-user computer used in a time-sharing mode in which users can share the hardware but with separate stand-alone application (Mallach 2002). DSS can also be developed using web-based services that may be widely used or may be used by multiple units of an organization are beginning to be offered as web-based DSS (Power and Sharda 2007). In addition, the web-based DSS expands the availability of operations and easy accessibility on internet without constraint of time and processing capabilities of the client machine ubiquitously with the additions of hyperlinks and external data/document sources over the stand-alone system (Power 2002).

DSS is a computer-based and is a comprehensive support system than other traditional techniques of decision-making for watershed management community who deal with semi-structured watershed problems such as surface/sub-surface water source long-term availability, hydrological, socio-economic and water quality issues. It is used for development of watershed management plans and operating rules for sustainable environment through policy making. Integration of GIS into spatial DSS system (SDSS) has given the researchers advantage for spatial analysis and visualization (Enache 1994). The ITC Netherlands had developed the first Integrated Land and Water Information System (ILWIS) in 1990s with the integration of GIS which is extremely helpful in spatial modeling. Adinarayana et al. (2006) designed a spatial decision support system for rural land use planning (SDSS/ LUP) to support decision making on area selection for different watershed management schemes for conservation planning by providing suggestions and hazard warnings for land use sustainability. Hellweger and Maidment (1999) developed an automated procedure in ARC/INFO and ARC/View to produce the connection of hydrologic elements using the geographic data, which was used to identify the hydrologic elements in Tenkiller reservoir watershed in Oklahoma, USA. In the context of rural development planning, GramyaVikas was developed as a web-based distributed collaboration model to assist the rural extension community in their own decision-making processes in a more interactive, integrated and coordinated manner (Adinarayana et al. 2008). The recent technologies such as Geospatial Information Communication Technology (GeoICT) and Wireless Sensor Network (WSN) were integrated to formulate Geo-Sense, a wed-based DSS to facilitate precision agriculture services (Sudharsan et al. 2012). Many tools for watershed analysis and management are being developed for integrated planning. The modern technology and thinking offered by the advent of the stand-alone or web-based DSS is highly complementary to assure the goals of watershed analysis through solving the complex decision-making process.

1.4 Need for Advanced and Augmented Techniques for Watershed Management

In the past, most of the studies consider the watershed management that consists of water resources planning for entire river basin/catchment/sub-catchment scale. The challenge on watershed management at macro/micro level is to make reliable assessment of water demand and availability with the given data. The major gap in the evolving watershed management concept at macro or micro level is due to very limited distribution and exchange of information and datasets caused by different norms, policies, institutional and organizational factors. Hence, it was found that, for studying the detailed aspects of soil and water management as well as to implement hydrological modeling techniques with adequate datasets, it is necessary to understand watershed management at micro level for decentralized planning. Also, land irrigability and capability classification for micro level watershed planning are needed so that the farmer can use better parcels for intensive cultivation with proper conservation measures and soil improving practices.

Technologies are available to solve many watershed problems (irrigation scheduling, water release cycles in canal command area, etc.). However, methods are further needed to effectively demonstrate the benefits of instituting environmentally sound watershed management programmes. Technologies need to be demonstrated in an effective way for easy accomplishment of planes and implementations by the users. These tools and techniques also need to be cost effective. Besides, the physically based hydrological models are very complex and have lots of input parameters and, as previously explained, the major problem is being related to availability of adequate database. Hence, viable methodology must be prevailed over to serve the novice user.

It was found that technologies alone are not producing the expected results to facilitate sustainable development and natural resource management. It is vital to carry out further studies, research and analysis on the concepts and approaches of watershed management. Studies are required on what has been accomplished with existing ones and how these can be made even better.

New concepts and approaches should be developed to reduce the rate of watershed degradation and to improve agriculture development. It is also true that the management and conservation of land and water resources can not be sustainable nor could they be replicated unless all the physico-chemical and biological processes including people's concerns are not taken into account. In addition, participatory approach for water resources conservation and management would be more helpful in reducing the rate of degradation of water resources.

In water resources management, previously specified case studies related to DSS and Information Systems are site/problem precise having focused only on specific hydrological processes such as runoff/erosion, etc. (Palmer and Holmes 1988; Reitsma 1996; Ito et al. 2001; Shim et al. 2002; Koutsoyiannis et al. 2003; Zhang et al. 2004; Mysiak et al. 2005; Calder et al. 2008; Tian-en et al. 2009; Alminana et al. 2010). Furthermore, a few watershed DSSs are merely related to environmental problems e.g. water quality assessment (Poch et al. 2004; Rao et al. 2007; Mullinix et al. 2009; Weng et al. 2010). As watershed management process not only include data related to spatial and temporal attributes but also includes data related with surface water storage, ground water recharge and ground water management, hydrology climatology, agriculture, topography, environmental and socio-economic aspects. The challenging task for natural resources researcher is to combine all the concepts and to prepare entire spatial and non-spatial attribute database by amalgamating the leading edge technologies to form decision making analysis technique. Thus, development of entire watershed management decision support and information system is needed to integrate the Agriculture–Water–Soil–Climate constituents to accomplish the natural resources management and, in turn, sustainable development. Also, the complexities of hydrological models cause many problems for the novice user, particularly in SAT region where spatio-temporal variability of environmental factors is high.

There is a need of appropriate modelling and application of modern techniques to integrate Agriculture–Water–Soil–Climate environments to optimize and allocate the land and water resources properly. Suitable measures, data, and modern techniques such as GIS, Remote Sensing and soft computing tools that could be utilized to manage watersheds imply appropriate technologies at the farmer level and provide watershed services for upstream and downstream areas.

Keeping the above points in view, an attempt has been made to develop a viable, full-fledged, user friendly and holistic toolkit for integrated watershed planning and management of its natural resources, christened "WATMIS: Watershed Management Information System". WATMIS is a web-based information/decision support system (DSS) that integrates soil-vegetation-climate-environment dynamics in formulation of hydrological systems water balance model through the multi-cropping agricultural practices with distinct irrigation practices.

2 Study Area

The WATMIS was demonstrated over a watershed located at Pimpalgaon Ujjaini in Ahmednagar District of Maharashtra state, India (Fig. 1). Maharashtra, has only 13 % of the cropped area under irrigated water, and the remaining area is dependent on rainfall only. In this state larger flow irrigation projects are working but it can not satisfy irrigation water need for all, particularly at rural community level. Therefore, planning these units; using recent tools and techniques, and in situ methods of soil and water conservation at micro watershed and/village level is necessary. The watershed is located between 74°45′00″E to 74°51′00″E longitude and 19°08′43″N to 19°11′31″N latitude.

The study area falls in transition zone between mountainous and water scarcity region of central plateau region and characterized by very shallow to very deep black, moderately permeable soils, and shows sandy loam to clay texture over nearly level (<1 %) to very steep sloping land (>35 %). Agriculture is practiced mainly with single agricultural season (*kharif*: June–October or *rabi*: November–March) with exception under double crop cultivation (Aher et al. 2012). The limiting soil moisture is the major threat influencing crop yield. The watershed represents water scarcity, acute drought prone and rain shadow conditions and falls under semi-arid tropical climatic conditions with the mean minimum and maximum temperatures of about 11 and 35 °C in winter and summer seasons, respectively, with an average rainfall of about 650 mm. Thereby, all these characteristics suggest water scarcity and land degradation are major environmental intimidation.

3 Conceptual Design

In a sound watershed management framework, various complex decision making processes are involved with structural and non-structural practices that can be undertaken to optimize of land and water resources, prevent soil erosion, stabilize water demand, and to increase productivity through efficient land use planning. The WATMIS attempt illustrate the development of a viable and generic toolkit for integrated watershed planning and management of its natural resources. The system is conglomeration of multiple technologies like Geographical Information System (GIS), Remote Sensing (RS), Global Positioning System (GPS), hydrological modelling, soft-computing tools, etc.

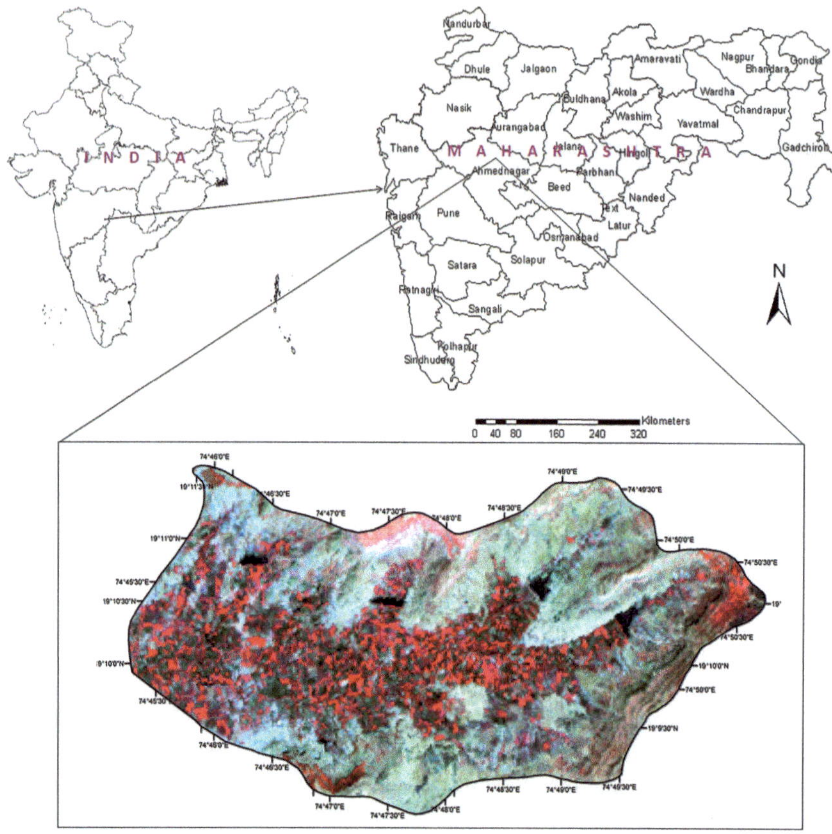

Fig. 1 Study area location map of Pimpalgaon Ujjaini watershed, Ahmednagar (MH), India

3.1 Data Used

Various spatial, non-spatial, temporal, attribute, and thematic datasets were used in WATMIS. Satellite Data of Landsat 7 Enhanced Thematic Matter (ETM/ETM+), particularly in the growing phases of crops (September–December), were used for obtaining land use distributions as well as irrigation water necessitate. The meteorological (Julian day of year, mean relative humidity, solar radiation, open pan evaporation, wind speed, and daily minimum and maximum air temperature), cropping system, soil (field capacity, soil type, permanent wilting point) and watershed datasets were used for dynamic hydrological modeling to obtain maximum crop yield through optimal allocation of watershed resources.

3.2 Tools and Technologies Used

- *PostgreSQL*: PostgreSQL (9.0) is an open source, standards compliant and highly customizable Object Relational Database Management System (ORDBMS) based on Relational Database Management Systems (RDBMS), developed by PostgreSQL Global Development Group. It provides support for foreign keys, joins, views, triggers, and stored procedures (in multiple languages) and includes most of the data types such as integer, numeric, boolean, char, variable character, date, interval, and timestamp (PostgreSQL, 9.0). GiST (Generalized Search Tree) with PostGIS public project supports geographic objects in PostgreSQL by allowing it to be used as a spatial database for GIS. ORDBMS has been used as back end for retrieval of attribute/non-spatial and spatial datasets for further analysis and modeling.
- *Apache server*: Apache server (2.2) developed by Apache Software Foundation, provides open-source web server platform for online distribution of website services (Apache 2013). In WATMIS, the server processes the requests received from user and sent back the generated information through the web browser.
- *Hyper Text Markup Language* (*HTML*): WATMIS uses Hyper Text Markup Language (HTML), an extended version of Standard Generalized Markup Language (SGML), for development of documents on the World Wide Web (WWW) (World Wide Web Consortium 2013). HTML expresses data input/display and environmental modelling results through user.
- *Hypertext Preprocessor* (*PHP*): Open source Hypertext Preprocessor (PHP), a server-side scripting language, in combination with HTML was used for WATMIS development. As it provides server-side scripting, command line scripting and desktop applications with a graphical user interface (GUI), which widely supports in executing dynamic, interactive database-server enabled web pages (The PHP Group 2012) on client/user action.
- *Map Server*: WATMIS uses Map Server Open Source geographic data rendering engine, developed by the University of Minnesota. Map Server WWW-GIS technology is built up on numerous Open Geospatial Consortium (OGC) standards and provides access to DBMS and also provides a user friendly interface for data input, query, analysis and display of various vector and raster datasets (MapServer 2013).
- *Java Script*: It is a programming language which can be combined into HTML pages for providing high level of interactivity to web pages than simple HTML (Oracle 2013). This language is used for dynamic visualization of data on internet.
- *pMapper*: *pMapper* framework was used for querying and visualization of spatial information on web. It has extensive functionality as well as multiple configurations to provide the setup of a MapServer application based on PHP/MapScript (pMapper 2013).

3.3 System Architecture

WATMIS is a user friendly-interactive-web based decision support system that consists of different hydrological processes and their modelling for sustainable development and management of natural resources on a watershed basis. The fundamental processes of hydrological cycle such as rainfall, surface runoff, ET, etc. along with various components of watershed management techniques were integrated and modeled for developing the WATMIS that can integrate and handle various models and GIS data sets. Besides, an integrated hydrological simulation model was developed and assimilated with user interfaces to form a holistic structure for watershed management decision making processes.

A schematic representation of the integrated WATMIS is depicted in Fig. 2. The system is implemented as a layered structure, with every layer corresponding to a different functionality. The design of WATMIS consists of following three layers:

- Database Management System (DBMS)
- Application layer (AL)
- User-Database-Model Interface (UDMI)

The Database Management System (DBMS) includes various spatial as well as non-spatial multi source (satellite, GPS, ground and map-based data and reports) datasets of the watershed such as soil, hydrologic, meteorological, geologic, land and vegetation. Server Side Application Layer (AL) is implemented using Apache server-2.2 (Apache 2013) which is a powerful and flexible technology for hosting dynamic web pages. This provides user-friendly and robust platform for various modelling environments through data assimilation-integration-analysis. The User-Database-Model Interface (UDMI) consists of different forms for receiving the information from the user by using HTML (Hyper Text Markup Language) and PHP (Hypertext Preprocessor) languages. UDMI presents robust mechanism for transfer/access of the data for user community. It also provides platform for data analysis and modelling of the hydrologic components for efficient watershed management.

4 Online Generation and Implementation of WATMIS

In SAT regions, the available water can be either surface water or sub-surface water (ground water) or both as a result of rainfall. Agricultural water requirement is a major component of water demand in a watershed. Also, the water demands can be domestic needs of human population as well as of animals. Particularly in semi-arid tropics, to get the desired benefits of watershed management approach, adoption of certain soil and water conservation measures in the watershed is essential. Hence, an attempt has been made for natural resources planning and management on a watershed basis. The watershed based natural resources planning

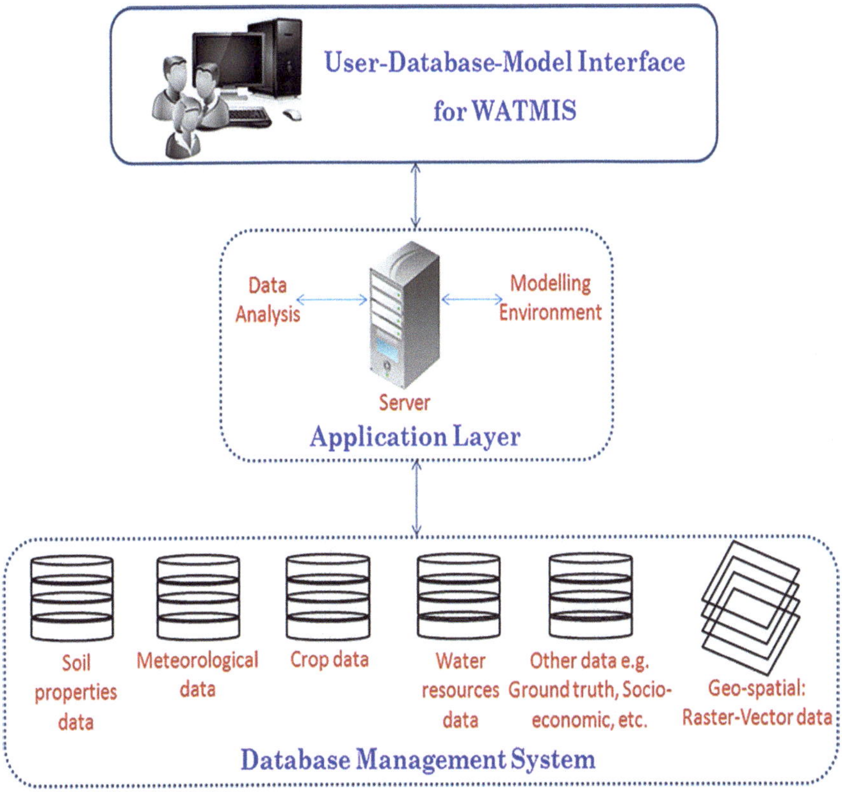

Fig. 2 System architecture of watershed management information system (WATMIS)

approach will basically consists of water availability assessment, modelling of major water availability-demand components, and operational planning of water resources by considering all the processes in hydrological cycle of a watershed. Ahmednagar district in Maharashtra, an inland and drought-prone district, falling under SAT region has been chosen in order to verify the management approaches and suggest the best management practices (BMPs) by demonstrating and developing a user friendly and widely supported open-source software based watershed management information system for better decision making.

WATMIS is designed and developed by using modern scientific developments and technologies to achieve holistic, sustainable watershed management aspects on the basis of architecture shown in Fig. 2. The system consists of various types of databases such as agricultural, meteorological, irrigation and soil as well as hydrological models and spatial/non-spatial user database model interface. The homepage of WATMIS (Fig. 3) depicts various information repository services about natural resources management information system with which the user communicates to the system by furnishing the login details. The user communication plays a significant role in endowing the BMPs. Therefore, the module

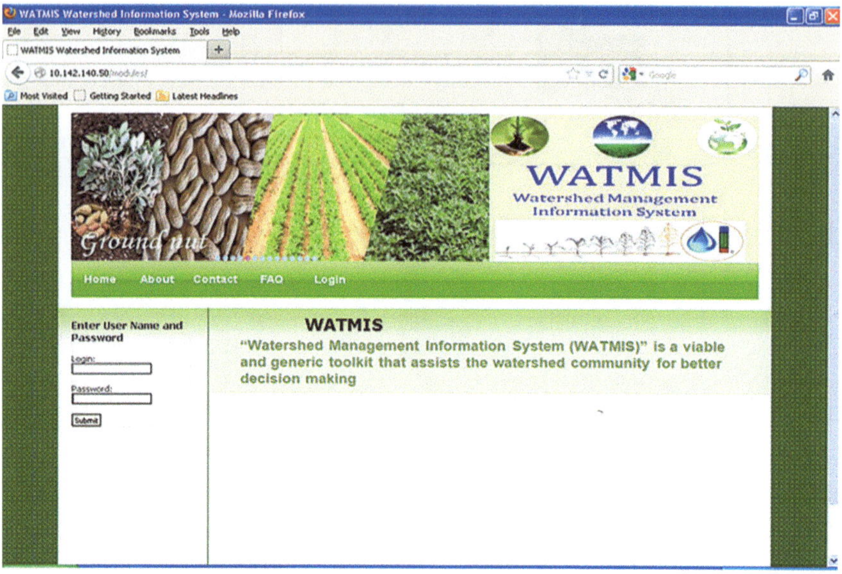

Fig. 3 Homepage of WATMIS

consists of learning repository services that provides introduction and basic information about the watershed modelling environment, Frequently-Asked-Questions (FAQ) and expertise contact details. Furthermore, through the login, user is navigated to simulation and modelling environment.

To increase the robustness and applicability of land and water resources management system, an integrated object-oriented relational approach based database is designed for development of various interfaces or modules. The data necessitated to be handled consists of multi scale spatial and non-spatial data from various reports, ancillary data sources, satellite, geographical, hydrological and environmental data sets such as, topographic maps, soil types, cropping system, meteorological data, etc. The DBMS layer is implemented using Apache server and is used for efficiently designing the tables, relationships, referential integrity rules and queries. All the datasets in DBMS have ability to data input, accessibility, update, visualization, and analysis through proper interaction among themselves via primary key and foreign key relationship.

WATMIS provides spatial database interaction system which is immense helpful in visualization of watershed framework layers such as land use patterns, soil types/texture, availability of nutrients, land capability and irrigability patterns, etc. The spatial dataset interface was designed by using open source GIS application, 'Map Server', in conjuction with front-end application 'p.mapper', which is implemented through incorporating Java Script, Map Script and PHP scripts. This facilitates for visualization and database query of the spatial location of interest directly for decision making and implementing the planning strategies. Furthermore, these

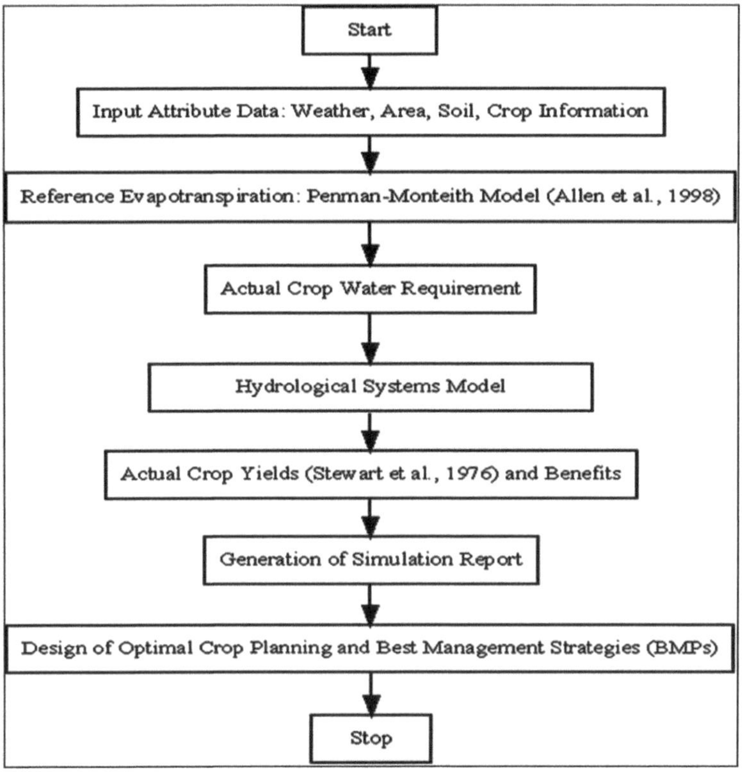

Fig. 4 Flow design for optimum allocation of available land and water resources

query modules provides the information about the topography, soils, crops and management zones of the sub-watersheds/parcel units on which soil and water resources conservation measures can be implemented through watershed prioritization.

The simulation UIL consists of various forms for accepting information from the user/client and validating those forms using PHP script. Privilege has been provided to user to select the number of crops, the soil system on which the crop is grown and the temporal variation of cropping season for which meteorological data such as relative humidity, rainfall, sun shine hours, daily maximum and minimum temperatures, wind speed, etc. was taken to run the hydrological model. The modelling approach adopted through integration of soil–vegetation–climate–environment dynamics in formulation of hydrological model for the multi-cropping agricultural systems (Fig. 4).

In data analysis and modelling environment, the basic input data can be retrieved through DBMS to calculate the crop water requirement (Fig. 5) and actual crop water demand, which further provided as an input for analyzing the hydrological model to obtain the actual crop yields and benefits with distinct

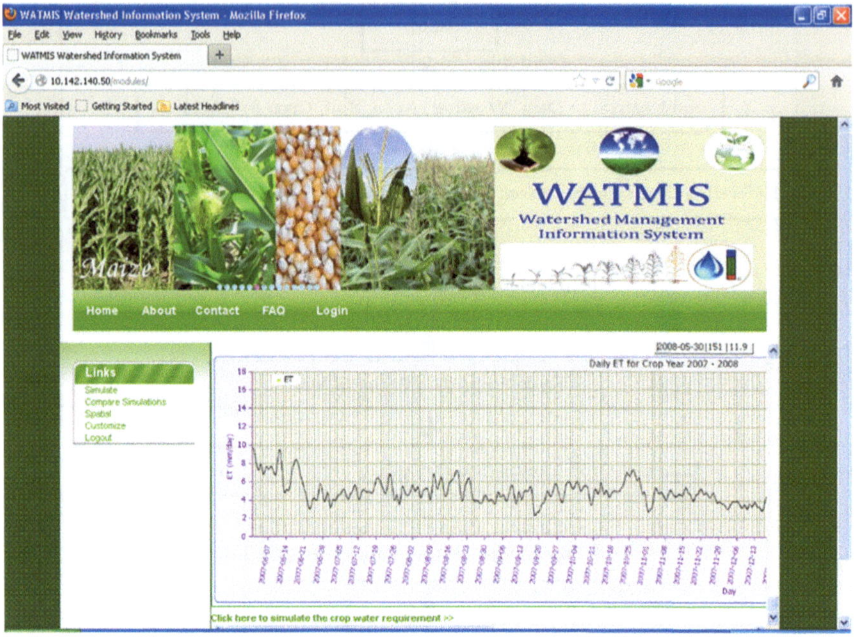

Fig. 5 Simulation of hydrological model

irrigation practices. Furthermore, simulation report generation service was also provided to the client that facilitates in comparative analysis among different possible planning strategies to accomplish the optimal solution. If the given irrigation strategy is not appropriate, the information service was provided to farmer about when to irrigate and how much to irrigate. Thus, the simulation emphasizes on factors influencing crop yields such as irrigation strategies, soil types, etc. to facilitate judicious planning and management of natural resources through optimization of available land and water properly.

During the course of operation of WATMIS, expert advices in the form of help menu or FAQ's provide sound basis for taking appropriate decisions. The 'Customize' operation facilitates user to modify, add, delete the input data for modelling the WATMIS for a given area of interest. Currently, web-enabled WATMIS is operational at local host and the system will be deployed at various user organizations with their broadband lines to make it online.

5 Conclusion

A viable, cost-effective, object-oriented and generic toolkit, called "WATMIS: Watershed Management Information System" using emerging tools and technologies such as, soft computing, GIS, RS, GPS, hydrological modelling, etc. was developed for online integrated watershed planning and management of its natural resources. The effective development of the WATMIS illustrates successful formulation of the framework for supported web-based sustainability.

WATMIS successfully assimilates Agriculture–Water–Soil–Climate continuum for attaining the suitable irrigation level for multiple cropping systems. In addition, it assesses crop yield, demarcating prioritization zones for soil and water conservation management and implementing knowledge repository services during information-decision support with spatial/non-spatial database management, visualization, analysis, query and user individualized customization utilities for optimization and management of land and water resources properly. With augmentation of the frequent temporal variation in satellite data, the system can be improved towards real to near-real time evaluation. The application of WATMIS will be useful to various stakeholders such as agriculturists, rural extension community, and water resources managers for better decisions making.

References

Adelman L (1992) Evaluating decision support and expert systems. John Wiley and Sons, New York

Adinarayana J, Laurenson M, Ninomiya S (2006) Web-based decision support system for rural land use planning-WebLUP-a prototype. Agric Eng Int CIGR E-J 8 (Manuscript IT 05 005)

Adinarayana J, Tewari G, Azmi S, Sudharsan D (2008) GramyaVikas—a distributed collaboration model for rural development planning. J Comput Electron Agric 62(2):128–140

Aher PD, Adinarayana J, Gorantiwar SD (2012) Use of morphological characteristics for multi-criteria evaluation through fuzzy analytical hierarchy process for prioritization of watersheds. In: 21st century watershed technology: improving water quality and the environment conference proceedings, Bari, Italy, pp 12–13639, 27 May–1 June 2012

Alminana M, Escudero LF, Landete M, Monge JF, Rabasa A, Soriano J (2010) WISCHE: a DSS for water irrigation scheduling. Omega 38(6):492–500

Apache (2013) Apache HTTP server project. http://httpd.apache.org/. Accessed on: 30 Jan 2013

Bashir MA, Hata T, Tanakamaru H, Abdelhadi AW, Tada A (2008) Satellite-based energy balance model to estimate seasonal evapotranspiration for irrigated sorghum: a case study from the Gezira scheme, Sudan. Hydrol Earth Syst Sci 12(4):1129–1139

Bryan BA, King D, Ward JR (2011) Modelling and mapping agricultural opportunity costs to guide landscape planning for natural resource management. Ecol Ind 11(1):199–208

Burkhard B, Kroll F, Nedkov S, Müller F (2012) Mapping ecosystem service supply, demand and budgets. Ecol Ind 21:17–29

Calder I, Gosain A, Rao MSRM, Batchelor C, Garratt J, Bishop E (2008) Watershed development in India. 2. New approaches for managing externalities and meeting sustainability requirements. Environ Dev Sustain 10(4):427–440

Conoscenti C, Agnesi V, Angileri S, Cappadonia C, Rotigliano E, Märker M (2013) A GIS-based approach for gully erosion susceptibility modelling: a test in Sicily, Italy. Environ Earth Sci 1–17. DOI: 10.1007/s12665-012-2205-y

Elhag M, Psilovikos A, Manakos I, Perakis K (2011) Application of the SEBS water balance model in estimating daily evapotranspiration and evaporative fraction from remote sensing data over the Nile delta. Water Resour Manage 25:2731–2742

Enache M (1994) Integrating GIS with DSS: a research agenda. In: Proceedings of Urban and Regional Information Association (URISA), Milwaukee, Wisconsin, pp 154–166, 12–15 Aug

Esteves TCJ, Kirkby MJ, Shakesby RA, Ferreira AJD, Soares JAA, Irvine BJ, Ferreira CSS, Coelho COA, Bento CPM, Carreiras MA (2012) Mitigating land degradation caused by wildfire: application of the PESERA model to fire-affected sites in central Portugal. Geoderma 191:40–50
Georgoussis H, Babajimopoulos C, Panoras A, Arampatzis G, Hatzigiannakis E, Ilias A, Papamichail D (2009) Regional scale irrigation scheduling using a mathematical model and GIS. Desalination 237(2):108–116
Hellweger FL, Maidment DR (1999) Definition and connection of hydrologic elements using geographic data. J Hydrol Eng 4(1):10–18
Ito K, Xu ZX, Jinno K, Kojiri T, Kawamura A (2001) Decision support system for surface water planning in river basin. J Water Resour Plann Manag 127(4):272–276
Jain MK (1996) Geomorphologic and land use planning for Danda watershed. National Institute of Hydrology, Roorkee
Jensen JR (2007) Remote sensing of the environment: an earth resource perspective, 2nd edn. Prentice Hall, New Jersey
Kim J, Noh J, Son K, Kim I (2012) Impacts of GIS data quality on determination of runoff and suspended sediments in the Imha watershed in Korea. Geosci J 16(2):181–192
Koutsoyiannis D, Karavokiros G, Efstratiadis A, Mamassis N, Koukouvinos A, Christofides A (2003) A decision support system for the management of the water resource system of Athens. Phys Chem Earth 25(14–15):599–609
Little JDC (1970) Models and managers: the concept of a decision calculus. Manage Sci 16(8-B):466–485
Liyantonoa Kato Tb, Kuroda Hc, Yoshida K (2013) GIS analysis of conjunctive water resource use in Nganjuk district, east Java, Indonesia. Paddy Water Environ 11(1–4):193–205
López-Vicente M, Poesen J, Navas A, Gaspar L (2013) Predicting runoff and sediment connectivity and soil erosion by water for different land use scenarios in the Spanish Pre-Pyrenees. Catena 102:62–73
Mahajan S, Panwar P (2005) Land use changes in Ashwani Khad watershed using GIS techniques. J Indian Soc Remote Sens 33(2):227–232
Mallach EG (2002) Decision support and data warehouse systems. Tata Mcgraw Hill, New Delhi
MapServer (2013) MapServer 6.2.0 documentation. http://mapserver.org . Accessed on: 02 Jan 2013
Mason DC, Cobby DM, Horritt MS, Bates PD (2003) Floodplain friction parameterization in two-dimensional river flood models using vegetation heights derived from airborne scanning laser altimetry. Hydrol Process 17:1711–1732
Mullinix C, Hearn P, Huajun Z, Aguinaldo J (2009) Web-based decision support and visualization tools for water quality management in the Chesapeake Bay watershed. In: The 17th international conference on geoinformatics, Fairfax, Virginia, USA, pp 1–6, 12–14 Aug 2009
Mysiak J, Giupponi C, Rosato P (2005) Towards the development of a decision support system for water resource management. Environ Model Softw 20(2):203–214
Nair ASK (2009) A new scientific management approach to water related natural disasters. In: Proceedings of Kerala environment congress, Thiruvanthapuram, pp 143–154, 19–21 Aug
Nahry AHE, Ali RR, Baroudy AAE (2011) An approach for precision farming under pivot irrigation system using remote sensing and GIS techniques. Agric Water Manag 98(4):517–531
Oracle (2013) Java scripting programmer's guide. http://docs.oracle.com . Accessed on: 27 Jan 2013
Palmer RN, Holmes KJ (1988) Operational guidance during droughts: expert system approach. J Water Resour Plann Manag 114(6):647–666
Park JH, Hur YT (2012) Development and application of GIS based K-DRUM for flood runoff simulation using radar rainfall. J Hydro-Environ Res 6(3):209–219
pMapper (2013) pMapper documentation. http://www.pmapper.net/ . Accessed on: 09 Jan 2013

Poch M, Comas J, Rodrý I, Sa M, Corte U (2004) Designing and building real environmental decision support systems. Environ Model Softw 19(9):857–873

Power DJ (2002) Decision support systems: concepts and resources for managers. Greenwood/ Quorum, Westport

Power DJ, Sharda R (2007) Model-driven decision support systems: concepts and research directions. Decis Support Syst 43:1044–1061

Rao M, Fan G, Thomas J, Cherian G, Chudiwale V, Awawdeh M (2007) A web-based GIS decision support system for managing and planning USDA's conservation reserve program (CRP). Environ Model Softw 22(9):1270–1280

Rashed M, Idris Y, Shaban M (2006) Integrative approach of GIS and remote sensing to represent the hydrogeological and hydrochemical conditions of Wadi Qena—Egypt. In: The 2nd international conference on water resources and arid environment, Saudi Arabia, pp 26–29, Nov 2006

Rawat PK, Tiwari PC, Pant CC, Sharama AK, Pant PD (2011) Modelling of stream run-off and sediment output for erosion hazard assessment in Lesser Himalaya: need for sustainable land use plan using remote sensing and GIS: a case study. Nat Hazards 59(3):1277–1297

Reitsma RF (1996) Structure and support of water-resources management and decision-making. J Hydrol 177(3–4):253–268

Roy UN (2005) Participatory RS and GIS for micro level watershed planning and management. In: 31st international symposium on remote sensing of environment, Saint Petersberg, 20–24 June 2005

Sabins FF Jr (1987) Remote sensing: principles and interpretation. W. H. Freeman and Co., New York

Saidi S, Bouri B, Dhia HB, Anselme B (2009) A GIS-based susceptibility indexing method for irrigation and drinking water management planning: application to Chebba-Mellouleche Aquifer, Tunisia. Agric Water Manag 96(12):1683–1690

Samarakoon L (2005) Basic geographic information system (GIS) vector/raster model. ISPRS workshop on remote sensing and GIS for watershed management, Laos, 2–5 Dec

Shim KC, Fontane DG, Labadie JW (2002) Spatial decision support system for integrated river basin flood control. J Water Resour Plann Manag 128(3):190–201

Shrivastava PK, Tripathi MP, Das SN (2004) Hydrological modeling of a small watershed using satellite data and GIS technique. J Indian Soc Remote Sens 32(2):145–157

Singh S (2010) Geography for UPSC prelims, 3rd edn. Tata McGraw-Hill Education, New Delhi

Srivastava VK (2003) Role of GIS in natural resources management. In: Thakur B (ed) Perspectives in resource management in developing countries. Concept Publishing Company, New Delhi, pp 479–484

Steinfeld CMM, Kingsford RT, Laffan SW (2013) Semi-automated GIS techniques for detecting floodplain earthworks. Hydrol Process 27:579–591

Strager MP, Fletcher JJ, Strager JM, Yuill CB, Eli RN, Todd PJ, Lamont SJ (2010) Watershed analysis with GIS: the watershed characterization and modeling system software application. Comput Geosci 36(7):970–976

Sudharsan D, Adinarayana J, Tripathy AK, Ninomiya S, Hirafuji M, Kiura T, Desai UB, Merchant SN, Reddy DR, Sreenivas G (2012) GeoSense: a multimode information and communication system. ISRN Sensor Networks, 2012, Article ID 215103, 13 p, doi: 10.5402/ 2012/215103

The PHP Group (2012) PHP manual. http://php.net/ . Accessed on: 14 Sept 2012

Tian-en C, Li-ping C, Yunbin G, Yanji W (2009) Spatial decision support system for precision farming based on GIS web service. In: Proceedings of information technology and applications, 2009 (IFITA '09), China, pp 372–376, 15–17 May

Vemu S, Pinnamaneni UB (2011) Estimation of spatial patterns of soil erosion using remote sensing and GIS: a case study of Indravati catchment. Nat Hazards 59(3):1299–1315

Vittala SS, Govindaiah S, Gowda HH (2008) Prioritization of sub-watersheds for sustainable development and management of natural resources: an integrated approach using remote sensing, GIS and socio-economic data. Curr Sci 95(3):345–354

Walsh MR (1993) Toward spatial decision support systems in water resources. J Water Resour Plann Manag 119(2):158–169

Weng SQ, Huang GH, Li YP (2010) An integrated scenario-based multi-criteria decision support system for water resources management and planning—a case study in the Haihe River Basin. Expert Syst Appl 37(12):8242–8254

World Wide Web Consortium (2013) World Wide Web Consortium (W3C). http://www.w3.org/. Accessed on: 12 Jan 2013

Zhang J, Li J, Zhu Y (2004) Integrating crop simulation models with WebGIS for remote crop production management. In: Proceedings of international geoscience and remote sensing symposium, Anchorage, Alaska, pp 2598–2600, 20–24 Sept

Sensitivity Exploration of SimSphere Land Surface Model Towards Its Use for Operational Products Development from Earth Observation Data

George P. Petropoulos, Hywel M. Griffiths, Pavlos Ioannou-Katidis and Prashant K. Srivastava

Abstract The use of Earth Observation (EO) data combined with land surface process models is at present being explored to assist in better understanding the natural processes of the Earth as well as how the different components of the Earth system interplay. However, before applying any modelling approach in performing any kind of analysis or operation, a variety of validatory tests needs to be executed to evaluate the adequacy of the developed "model" in terms of its ability to reproduce the desired mechanisms with the necessary reality. Sensitivity analysis (SA) is an integral and important validatory check of a computer simulation model or modelling approach before it is used in performing any kind of analysis operation. The present study builds on previous works conducted by the authors in which a sophisticated, cutting edge SA method adopting Bayesian theory has been implemented on a land surface process model called SimSphere with the aim of further extending our understanding of its structure and of establishing its coherence. This land surface model has been widely used as an educational tool in different Universities across the world, as a stand-alone tool and synergistically with EO data in deriving key parameters characterising land surface processes. SimSphere use is currently under investigation by two Space Agencies for deriving spatio-temporal estimates of energy fluxes and soil surface moisture from a technique in which the model is used synergistically with Earth Observation data. The GSA method employed here provided a further insight into the model's architectural structure, and allowed us to determine which model input parameters and parameter interactions exert a significant influence on the selected model outputs, and which are inconsequential. Analysis of the SA results indicated that only a small fraction of the model input parameters have an appreciable influence

G. P. Petropoulos (✉) · H. M. Griffiths · P. Ioannou-Katidis
Department of Geography and Earth Sciences, University of Aberystwyth, King Street, Aberystwyth, Wales SY23 2DB, UK
e-mail: george.petropoulos@aber.ac.uk

P. K. Srivastava
Department of Civil Engineering, University of Bristol, Bristol BS8 1TR, UK

P. K. Srivastava et al. (eds.), *Remote Sensing Applications in Environmental Research*, Society of Earth Scientists Series, DOI: 10.1007/978-3-319-05906-8_3,

on the examined target quantities. Results, however, did suggest the presence of highly complex interactions structure within SimSphere, which drove a considerable fraction of the variance of the variables simulated by the model. The main findings are discussed in the context of the future model use including its synergy with EO data for deriving the operational development of key land surface parameters from space.

Keywords Earth observation · Soil vegetation atmosphere transfer models · SimSphere · Energy fluxes · Sensitivity analysis · BACCO GEM-SA · Gaussian process emulators

1 Introduction

Understanding the natural processes of the Earth as well as how the different components of the Earth system interplay, especially in the context of global climate change, has been recognised by the global scientific community as a very urgent and important research direction requiring further investigation (Battrick et al. 2006). Recent investigations have focused on the combined use of remote sensing data and simulation process models in studying the Earth system and deriving physical parameters characterising land surface interaction processes (see reviews Olioso 1992; Petropoulos 2013). As a result, a wide range of methods have been proposed exploiting the use of different simulation process models with remote sensing data acquired in all regions of the electromagnetic radiation spectrum. These techniques essentially aim to improve estimates of key parameters characterising land surface processes by combining the horizontal coverage and spectrally rich content of remote sensing data with the vertical coverage and high temporal continuity of simulation process models. A review of such methods can be found in Moradkhani (2008) for example.

In particular, Petropoulos et al. (2009a) underlined the significant prospect of the methods which are based on combining the biophysical properties encapsulated in a satellite-derived scatter-plot developed between the surface temperature (Ts) and vegetation index (VI) maps with a Soil Vegetation Atmosphere Transfer (SVAT) model, named SimSphere, for deriving spatio-temporal estimates of latent (LE) and sensible (H) heat fluxes as well as of surface soil surface moisture. Indeed, their potential is also evidenced by the fact that several variants of it are currently used or being developed for deriving operational global estimates of key parameters characterising the Earth's energy and water budget, namely LE and H fluxes and/or soil surface moisture (Chauhan et al. 2003; Piles et al. 2011; Petropoulos and Carlson 2013). A description of the method workings can be found in Carlson (2007) and Petropoulos and Carlson (2011).

A variant of the "triangle" is presently under investigation for developing an operational scale global surface soil moisture maps by the National Polar-orbiting

Operational Environmental Satellite System (NPOESS), starting from 2016 (Chauhan et al. 2003). In addition, another variant of the "triangle" is currently explored in the framework of the PROgRESSIon project funded by the European Space Agency (ESA, ESA STSE Report, 2011). In addition to this, SimSphere model itself use is continually expanding worldwide both as an educational tool in universities and as a research tool where it is used either as a stand-alone application or synergistically with Earth Observation (EO) data to conduct studies aiming to improve understanding of land surface processes and their interactions.

Considering the research and practical work described above with respect to SimSphere use, particularly from the perspective of the global development of operational EO-based products, it is evidently of paramount importance to perform an all-inclusive validation of the model performance and to establish its coherence and reliability for simulating natural processes. In particular, the importance of including sensitivity analysis (SA) in any all-inclusive validation of modelling schemes planned to be used on a global operational scale cannot be overstated.

Petropoulos et al. (2009b) in an all-inclusive review of the SimSphere exploitation underlined the requirement to perform further validation exercises of the model operation in a wide range of implementation conditions as well as the importance of SA experiments. In response to this requirement, Petropoulos et al. (2009c, 2010) performed advanced SA on SimSphere based on a Gaussian process emulator. Their results provided for first time an insight into the model architecture and allowed establishing detailed mapping of the sensitivity of key target quantities simulated by the model in respect to its inputs. However, these SA studies were focused on particular variables simulated by SimSphere rather than a wide range of parameters. Thus, there is a need to verify the sensitivity of other variables simulated by the model, which are also of key importance to land surface interaction processes. In this context, the objective of this study was to build on previous SA works conducted on the model, exploring further the sensitivity of some new variables simulated by SimSphere.

2 Sensitivity Analysis: An Overview

SA can be defined as the process of determining the effect of changing the value of one or more input variables and observing the effect that this has on the model output. SA can generally help to understand the behaviour of a model, its coherence and correspondence to the real world (Saltelli et al. 2000; Gatelli et al. 2009), and to verify that the model concept corresponds to the natural system's behaviour in an appropriate manner (Nossent et al. 2011). In addition, SA can assist in establishing the dependency of the model outputs on its input parameters in how different parts of the model interplay, as well as to identify possible region(s) in the space of model input parameters that correspond to certain characteristics of the model prediction (Chen et al. 2012). As a result, SA provides a valuable method to identify critical input parameters and rank them in order of importance (Chen et al. 2010), offering guidance to the design of experimental programs as well as to more efficient model coding or calibration. This is because by means of a SA, irrelevant parts of the model may be dropped or a simpler model can be built or extracted from a more complex one (so-called 'model lumping'), reducing, in some cases significantly, the required computing power (e.g. Holvoet et al. 2005). GSA methods, despite their high computational demands, have become popular in environmental modelling due to their ability to incorporate parameter interactions and the relatively straightforward interpretation (Nossent et al. 2011).

According to Saltelli et al. (2000), SA methods are classified into three groups: screening methods, local sensitivity analysis (LSA) and global sensitivity analysis (GSA) methods. The screening methods are attractive because of their simplicity and their low computational burden, yet are only able to provide qualitative results. In LSA methods, the sensitivity of specific model outputs to the model inputs has been examined by varying the selected inputs around their expected 'nominal' values and observing the effect on the desired model output often by calculating the derivative of the output with respect to the input considered. Whilst useful, LSA methods have certain limitations such as the fact that derivatives are evaluated at the central estimate (meaning that results could be quite different at other points nearby) and also those LSA methods do not capture interactions between input parameters.

Global Sensitivity Analysis (GSA) methods apportion the output variability to the variability of the input parameters when they vary over their whole uncertainty domain, generally described using probability densities assigned to the model's input parameters. In GSA methods, the sensitivity of the input parameters is practically examined based on the generation of samples distributed across the parameter domain of interest. GSA methods have certain key advantages over LSA methods. These include their ability to incorporate the influence of the input parameters over their whole range of variation, thus providing results that are independent of any "modelers' prejudice", and results that are not expected to be site-specific (e.g. Saltelli et al. 1999). In addition, GSA methods are able to provide quantitative estimates not only of the most sensitive model inputs, but also of interactions between model input parameters (e.g. Schwieger 2004), yielding quantitative information on the degree of complexity of the model input–output relationships. GSA methods have been implemented to perform SA on different types of models in a large array of disciplines (e.g. Song et al. 2012; Feyissa et al. 2012).

The general procedure for obtaining sensitivity measures using GSA methods follows the steps described below (e.g. Saltelli et al. 2000; Schwieger 2004): (1) Define the probability distributions for each of the model input parameters, (2) Generate a training sample from the previously defined input probability distributions, (3) Evaluate the model output of interest (i.e. target quantity) using the generated sample, (4) Analyse the output variance and quantitatively estimate the contribution of each of the model inputs to the model output. The main approach used to perform Step 4 is based on a direct decomposition of the model output variance into factorial terms, called 'importance measures' (e.g. Ratto et al. 2001):

$$V(Y) = \sum_{i=1}^{s} D_i + \sum_{i \triangleleft j} D_{ij} + \cdots + D_{1 \ldots s} \tag{1}$$

$$D_i = V(E(Y|X_i)) \tag{1a}$$

$$D_{ij} = V(E(Y|X_i, X_j)) - V(E(Y|X_i)) - V(E(Y|X_j)) \tag{1b}$$

where

s denotes the number of inputs (so-called 'factors')
$V(Y)$ is the total variance of the output variable Y
D_i is the importance measure for input X_i
D_{ij} is the importance measure for the interaction between inputs X_i and X_j
$D_{1 \ldots s}$ denote analogous formulae for the higher order terms
$E(Y|X_i|)$ is the conditional expectation of Y given a value of *Xi* and the variance of $E(Y|X_i)$ is taken over all inputs factors which are fixed in the conditional expectations. This decomposition of variance into main effects and interactions is commonly known as ANOVA-HDMR, i.e. Analysis of Variance-High Dimensional Model Representation.

Sensitivity indices are computed by dividing the importance measures from Eq. 1 by the total output variance as follows:

$$S_i = \frac{D_i}{V(Y)}, \quad S_{ij} = \frac{D_{ij}}{V(Y)}, \tag{2}$$

These ratios S_i for $i = 1, \ldots, s$ are called correlation ratios, main effects or first order sensitivity indices, because each S_i delivers a direct measure of the share of the output variance explained by X_i. The main effect or first order sensitivity index S_i represents the expected amount of variance that would be removed from the total output variance if we knew the true value of X_i (within its uncertainty range). Thus, this measure quantifies the relative importance of an individual input variable X_i, in driving the total output uncertainty, and indicates where to direct future efforts to reduce that uncertainty. There are many different estimation methods for S_i, such as the Fourier Amplitude Sensitivity Test (FAST, Saltelli et al. 1999), Sobol's (1976) and others. Using similar formulae higher order sensitivity indices (joint effect indices) can be computed to quantify the sensitivity of the model output to input parameters interactions. However, in practice, because the estimation of S_i or S_{ij} or higher order is computationally very demanding and can be impractical, the SA is rarely carried out further after the computation of first order interaction indices (i.e. the second term of Eq. 1 above).

Thus, from the definitions of the above indices, and assuming non-correlated inputs, a complete series development of the output variance can be achieved:

$$\sum_{i} S_i + \sum_{i \lhd j} S_{ij} + \sum_{i \lhd j \lhd m} S_{ijm} + \cdots + S_{12\ldots k} = 1 \quad (3)$$

where higher order indices are defined in a similar way to Eq. 2.

In addition to the above indices, another measure, the total sensitivity index, is used to provide a cheaper computational method of investigating the higher order sensitivity effects as it collects all the interactions involving X_i in one single term. The total sensitivity index of a given factor X_i takes into account the main effect and the effect of all its interactions with other model inputs, and is defined as:

$$ST_i = \frac{D_i + D_{i,\sim i}}{V(Y)} \quad (4)$$

$D_{i,\sim i}$ indicates all interactions between factor X_i and all the others ($X_{\sim i}$).

The total sensitivity index represents the expected amount of output variance that would remain unexplained (residual variance) if only X_i were left free to vary over its range, the value of all other variables being known. The usefulness of the ST_i is that it is possible to compute them without necessarily evaluating the single indices S_i (and higher order ones), making the analysis computationally affordable. The total sensitivity indices are generally used to identify unessential variables (i.e. those that have no importance neither singularly nor in combination with others) while building a model. The existence of large total effects relative to main effects implies the presence of interactions among model inputs.

3 Materials and Methods

3.1 SimSphere Model

SimSphere was originally developed by Carlson and Boland (1978) and considerably modified to its current state by Gillies et al. (1997) and Petropoulos et al. (2013b). It is a Surface–Vegetation–Atmospheric Transfer (SVAT) model. SVAT models form a special category of deterministic simulation process models attempt to describe the physical processes that control energy and mass transfers in the soil/vegetation/atmosphere continuum (radiative, turbulent and water transfers) and provide estimates of the time course of soil and vegetation state variables with a fine time step (Olioso et al. 1999).

Figure 1 illustrates the different facets of the model's structure. SimSphere is a one-dimensional boundary layer model with a plant component implicitly referring to a horizontal area of undefined size that can be composed of a mixture of bare soil and vegetation. The underlying constraint in the model is taken as the balance

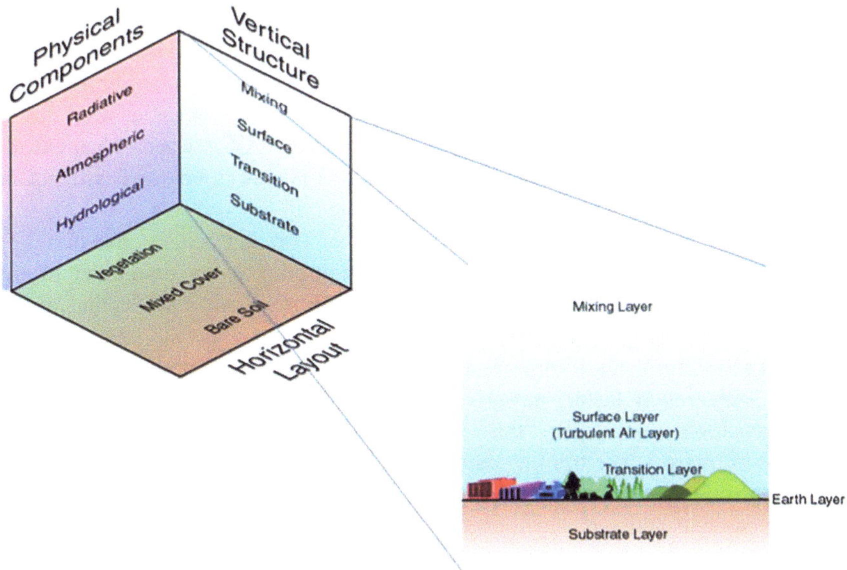

Fig. 1 The different facets of the SVAT model architecture (*left*) and a representation of the vertical structure of the plant-canopy model layers (*right*). Figure adopted from SimSphere user manual (http://www.aber.ac.uk/simsphere)

between all the energy fluxes at the Earth's surface. SimSphere is able to simulate the various physical processes that take place as a function of time in a column that extends from the root zone below the soil surface up to a level higher than the surface canopy. It performs simulations over a 24-h cycle at a time step of 15 or better, starting from a set of initial conditions given in the early morning (at 0600 h local time) with a continuous evolving interaction between soil, plant and atmosphere layers. The model simulates a number of parameters including the surface energy (H, LE and soil heat) fluxes at the soil surface and in, around and above the vegetation canopy, the flux of carbon dioxide between the atmosphere and the plants, the grid-cell scale surface temperature of the vegetation and soil mixture, as well as the temperature and humidity of the air above the surface canopy. An overview of SimSphere use can be found in Petropoulos et al. (2009b). The model is at present distributed globally from University of Aberystwyth (http://www.aber.ac.uk/simsphere).

3.2 The Bayesian GSA Method

In the present study, GSA was performed using a novel and sophisticated method based on Bayesian Analysis of Computer Code Outputs (BACCO; Kennedy and O'Hagan 2001). This also ensured consistency and comparability to previous SA works on SimSphere. Briefly, the BACCO method is based on the use of the Gaussian Emulation Machine for Sensitivity Analysis (GEM-SA) tool, the development of which was funded by the National Environmental Research Council, UK. Details concerning the statistical emulation process as well as the workings of the method and its advantages over other GSA approaches can be found elsewhere (Kennedy and O'Hagan 2000; O'Hagan 2006) and will not be provided here for brevity.

Briefly, the BACCO approach is composed of two key stages: the first step involves building a statistically-based representation (i.e. an emulator) of the model from a set of training data points derived from runs of the actual model under study, with these ideally designed to cover the multidimensional input space using a space-filling algorithm. The second stage uses the emulator (which is typically very much faster, easier and more efficient to run across the entire multidimensional input space than is the original model) to compute the SA quantities of interest. The training data are thus obtained from a set of actual model simulation runs, conducted for a specific set of initial conditions according to the sampling method adopted.

In the BACCO method, starting from a prior belief about the code (i.e. that it has no numerical error) and based on a GP model, Bayes' theorem and a set of the model code runs are used together to refine this prior information in order to yield the posterior distribution of the output, which is known as the emulator. The key prior assumption in emulator building using the GEM SA method is that the output is a reasonably smooth function of its inputs. The emulator calculates a mean function, which passes exactly through the observed runs and also quantifies the remaining uncertainty due to the emulator being an approximation to the true code. Cross-validation statistical measures can be generated automatically when the emulator is built in order to check the accuracy of both types of output. In addition to cross-validation, the emulator provides estimates of the smoothness of each of the model inputs, so-called "roughness values" which essentially describe how rapidly the output responds to changes in each input. The basic SA output from GEM SA includes the computation of the main and joint effects (pairwise interactions only) of the input parameters, as well as of the total effects (defined below). In addition, GEM SA provides a set of main effects plots, which are calculated for each of the model inputs. The percentage variance contribution of each input's main effect is also reported, providing a simple means of ranking the inputs in terms of their importance. The percentage variance component associated with each input measures the amount its main effect contributes to the total output variance, based on the uncertainty distributions for all inputs. It should be noted that, in general, summing the main effect contributions will not total to 100 % because of the additional contributions from the interaction effects. However, the total can be used to determine the degree of interactions.

The BACCO method has already supplied useful insights in various disciplines and a range of SA studies have underlined the advantages of this approach (Kennedy and O'Hagan 2001; Johnson et al. 2011; Kennedy et al. 2012; Parry et al. 2012). Petropoulos et al. (2009c) demonstrated for the first time the use of the BACCO method in performing a SA on SimSphere, providing an insight into the model structure. Petropoulos et al. (2010) performed a comparative study of various widely used process emulators including GEM and investigated the effect of sampling method and size on the sensitivity of key target quantities simulated by SimSphere.

3.3 BACCO Implementation

An overview of the SA study implemented in our study is illustrated in Fig. 2. Using the BACCO GEM, a SA was conducted to SimSphere on evaluating the sensitivity of key model outputs previously analysed in other SA studies of the model, namely:

- Daily Average Shortwave Incoming Radiation $\left(\overline{Rg_{daily}}\right)$,
- Daily Average Long-wave Downwelling Radiation $\left(\overline{Ldown_{daily}}\right)$,
- Daily Average Long-wave Upwelling Radiation $\left(\overline{Lup_{daily}}\right)$,
- Daily Average Soil Heat flux $\left(\overline{G_{daily}}\right)$,
- Daily Average Air Temperature at 1.3 m $\left(\overline{Tair_{1.3\mathrm{m}_{daily}}}\right)$ and
- Daily Average Air Temperature at 50 m $\left(\overline{Tair_{50\mathrm{m}_{daily}}}\right)$.

The sensitivity of these parameters was selected due to their importance in characterising land surface interaction processes and their relevance to the "triangle" techniques when the model is linked to EO data. In order to ensure consistency and comparability with Petropoulos et al. (2009c), BACCO GEM SA was implemented following the same methodology. The same design space of 400 SimSphere simulations developed using the LP-tau sampling method was used. In creating the input space from the 400 model runs, all SimSphere input parameters had been allowed to vary, except those of the geographical location (latitude/longitude) and atmospheric profile, for which a priori values were used taken from real observations from the Borgo Cioffi Italian CarboEurope site (Italy (40° 31′ 25.5″N, 14° 57′ 26.8″E) available for the 17th of November, 2004. All 30 model inputs were assumed to be uniformly distributed, with their probability distribution functions (PDFs) defined using the mean and variance taken from the entire possible theoretical range in SimSphere. No account was taken for possible co-variation between the parameters in the parameter set. Emulator performance was also carried out in order to verify the accuracy of the emulator fit to the training data using the cross validation method, based on a subset of the training data (i.e. the "leave final 20 % out" method). This method was selected because it provides more objective indication on the emulator performance as already explained.

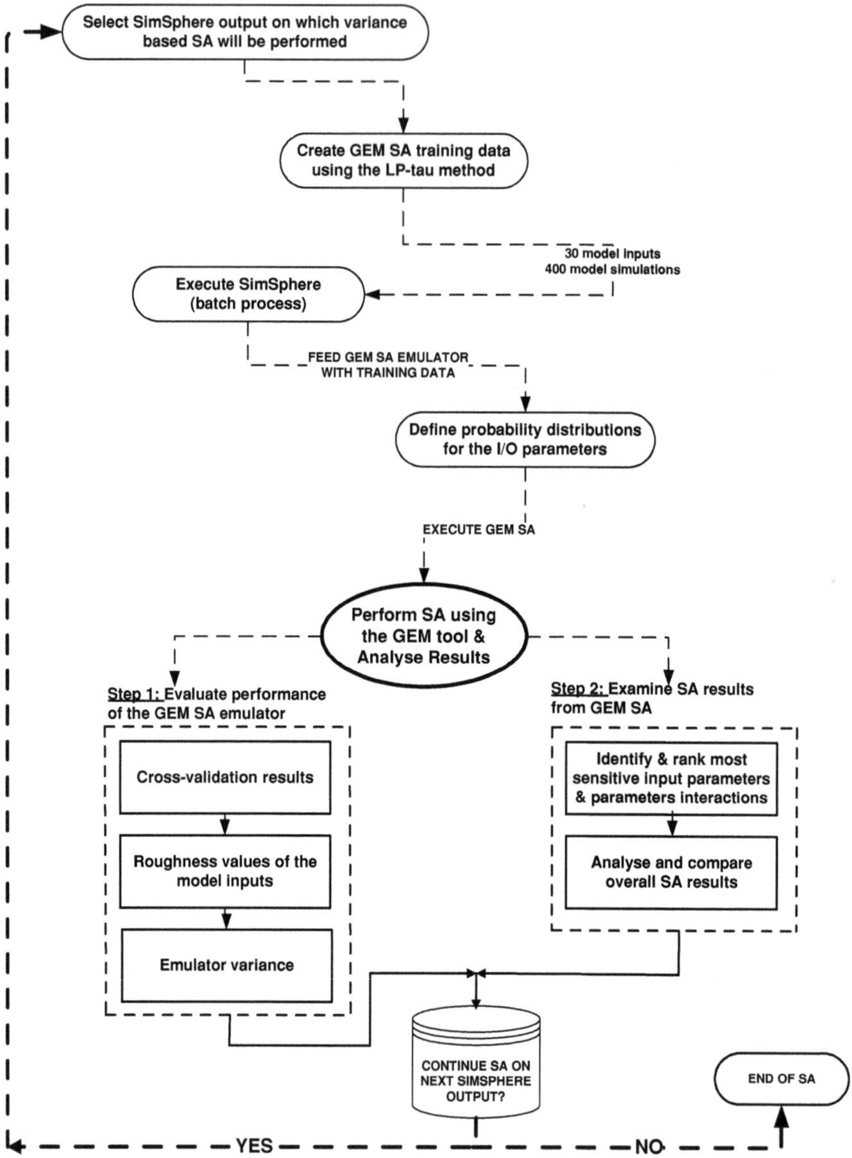

Fig. 2 Flowchart of the overall methodology followed in conducting the GSA on SimSphere using the BACCO method and the GEM SA software platform

4 Results

4.1 Emulator Validation

A number of statistical measures were used to quantify the uncertainty of the SA due to the emulator approximation were computed internally by GEM SA. These included the "cross validation root mean square error", the "cross-validation root mean squared relative error" and the "cross-validation root mean squared standardized error". In addition to these statistics, the emulator provides estimates of the smoothness of each of the model inputs, so called "roughness values" which essentially describe how rapidly the output responds to changes in each input. Each input has its own roughness and the (unitless) value of each roughness indicates the degree of non-linearity displayed by the effect of each input parameter. Finally, the "sigma-squared" statistical parameter is also computed in GEM SA from the validation sample used to statistically appreciate the performance of the emulator build. Within the BACCO GEM SA method this parameter is the variance of the emulator after standardising the output, and effectively provides a measure of the quality of the fit of the emulator to the original model code and should thus be as low as possible.

Tables 1 and 2 summarise the statistics concerning the evaluation of the emulator accuracy. As Table 2 shows, the roughness values for most of the input parameters are very low suggesting that the emulator is a very good approximation of the true model. For stem resistance, for example, roughness values for $\overline{Ldown_{daily}}$, $\overline{Lup_{daily}}$ and $\overline{Tair_{50\mathrm{m}_{daily}}}$ are 0, while those for $\overline{Rg_{daily}}$, $\overline{Tair_{1.3\mathrm{m}_{daily}}}$ and $\overline{G_{daily}}$ are 0.021, 0.003 and 0.007, respectively. For $\overline{Rg_{daily}}$, roughness values of model inputs ranged from 0 to 19.271, from 0 to 8.846 for $\overline{Ldown_{daily}}$, from 0 to 9.076 for $\overline{Lup_{daily}}$, from 0 to 9.192 for $\overline{Tair_{50m_{daily}}}$ from 0 to 6.473 for $\overline{Tair_{1.3m_{daily}}}$ and from 0 to 5.965 for $\overline{G_{daily}}$. Most roughness values obtained are less than 1.0, indicating that the emulator responded smoothly to input variation. The rare roughness values greater than 1.0 are italicized in Table 1 (including all roughness values for Aspect) and indicated some non-linearity in relationships between inputs and outputs but are not large enough to indicate extreme non-linearity. Sigma squared values, cross validation root mean squared-error and RMSE values shown in Table 2 are all reasonable and are comparable to values reported in previous studies (e.g. Petropoulos et al. 2009c, 2010) indicating moderate non-linearity and a good level of emulator accuracy.

4.2 SA Results

Table 3 shows the relative sensitivity of the model input parameters with respect to the sensitivity of $\overline{Rg_{daily}}$, $\overline{Ldown_{daily}}$, $\overline{Lup_{daily}}$, $\overline{Tair_{50\mathrm{m}_{daily}}}$, $\overline{Tair_{1.3\mathrm{m}_{daily}}}$ and $\overline{G_{daily}}$,

Table 1 Summarised statistics concerning the emulator accuracy evaluation for the SimSphere model outputs their sensitivity of which was examined. Input parameters withmain effect > 1 % and/or total effect > 1% are highlighted in grey

Model input		Roughness					
		Rg_{daily}	$Ldown_{daily}$	Lup_{daily}	$Tair_{50m_{daily}}$	$Tair_{1.3m_{daily}}$	G_{daily}
X1	Slope	1.504	0.949	0.856	1.584	0.631	0.622
X2	Aspect	19.271	8.846	9.076	9.192	6.473	5.965
X3	Station height	0.000	0.004	0.040	0.008	0.029	0.000
X4	Fractional vegetation cover	0.166	1.599	0.376	1.122	0.006	1.796
X5	LAI	0.137	0.111	0.165	0.297	0.536	0.465
X6	Foliage emissivity	0.003	0.008	0.011	0.015	0.000	0.077
X7	[Ca]	0.000	0.048	0.000	0.037	0.062	0.056
X8	[Ci]	0.000	0.017	0.307	0.015	0.042	0.001
X9	[03] in the air	0.000	0.004	0.022	0.000	0.025	0.055
X10	Vegetation height	0.000	1.435	0.035	1.288	2.457	0.397
X11	Leaf width	0.000	0.229	0.000	0.071	0.003	0.058
X12	Minimum stomatal resistance	0.000	0.108	0.042	0.039	0.084	0.007
X13	Cuticle resistance	0.000	0.028	0.000	0.057	0.022	0.446
X14	Critical leaf water potential	0.005	0.071	0.001	0.124	0.007	0.006
X15	Critical solar parameter	0.000	0.031	0.011	0.018	0.072	0.028
X16	Stem resistance	0.021	0.000	0.000	0.000	0.003	0.007
X17	Surface moisture availability	0.000	0.301	1.498	0.279	0.478	0.240
X18	Root zone moisture availability	0.000	0.003	0.034	0.000	0.029	0.005
X19	Substrate max. volum. water content	0.000	0.008	0.006	0.000	0.008	0.032
X20	Substrate climatological mean temper.	0.000	0.060	0.041	0.047	0.014	0.288
X21	Thermal inertia	0.000	0.000	0.030	0.000	0.000	0.824
X22	Ground emissivity	0.000	0.008	0.160	0.005	0.000	0.119
X23	Atmospheric precipitable water	0.000	0.171	0.099	0.078	0.013	0.076
X24	Surface roughness	0.000	1.062	0.109	0.976	4.200	0.129
X25	Obstacle height	0.000	0.000	0.052	0.000	0.010	0.000
X26	Fractional cloud cover	0.000	0.136	0.247	0.012	0.000	0.004
X27	RKS	2.081	0.000	0.005	0.000	0.020	0.000
X28	*CosbyB*	2.370	0.096	0.029	0.092	0.074	0.008
X29	*THM*	0.407	0.026	0.039	0.657	0.022	0.007
X30	*PSI*	0.876	0.000	1.615	0.000	0.037	0.020

Input parameters with main effect >1 % and/or total effect >1 % are italicized. Rows X1–X30 show roughness values for the different model outputs examined

with input parameters sensitive at >1 % main effect and/or >1 % total effect highlighted in italic. These data are also displayed on main effects plots in Fig. 3, whereas the decomposition of variance (main effects only) of the model inputs for the different "groups" of model outputs is illustrated in Fig. 4. There is a large range in both main effect values (<0.01–69.737, 37.101, 44.297, 35.859, 39.690

Table 2 Additional statistics concerning the emulator performance for the various SA tests conducted in our study

	$\overline{Rg_{daily}}$	$\overline{Ldown_{daily}}$	$\overline{Lup_{daily}}$	$\overline{Tair_{50m_{daily}}}$	$\overline{Tair_{50m_{daily}}}$	$\overline{G_{daily}}$
Fitted model parameters						
Sigma-squared	0.218	0.930	12.616	0.998	1.398	1.055
Cross-validation results						
Cross-validation root mean squared-error (Wm^{-2})	14.173	2.776	3.000	0.855	5.078	34.727
Cross-validation root mean squared relative error (%)	3.011	1.009	1.318	5.678	24.274	6.28E+01
Cross-validation root mean squared standardised error	0.825	1.369	4.395	1.270	1.606	1.271

and 36.344 % respectively) and total effect values (<0.01–85.315, 57.045, 64.585, 56.144, 60.073 and 56.249 %, respectively). The contribution of variance to the main effects is dominated by a small number of parameters. For example, slope and aspect are two of the three largest contributors of variance to all the model outputs examined in terms of main effects and total effects. For the main effects of $\overline{Rg_{daily}}$ they contributed 13.150 and 69.737 % respectively, 7.412 and 37.101 % for $\overline{Ldown_{daily}}$, 9.423 and 44.297 % for $\overline{Lup_{daily}}$, 8.351 and 35.859 % for $\overline{Tair_{50m_{daily}}}$ 9.738 and 39.690 % for $\overline{Tair_{1.3m_{daily}}}$ and 7.853 and 36.344 % for $\overline{G_{daily}}$. Aspect is, in fact, the largest contributor of percentage variance to main effects and total effects. Fractional vegetation cover contributed 14.910 % of main effects variance and 21.743 % to total effects variance to $\overline{Ldown_{daily}}$, 15.137 and 21.318 % to main and total effects variance for $\overline{Tair_{50m_{daily}}}$ and 10.611 and 17.274 % to main and total effects for $\overline{G_{daily}}$. Similarly, vegetation height is a significant parameter, contributing 3.366 % to the main effects of $\overline{Ldown_{daily}}$ and 3.139 % to the main effects of $\overline{Tair_{50m_{daily}}}$ and 10.084 % to the main effects of $\overline{Tair_{1.3m_{daily}}}$. Other important parameters include the surface moisture availability and surface roughness. The former contributes 6.515 % to the main effects of $\overline{Ldown_{daily}}$, 6.760 % to the main effects of $\overline{Tair_{50m_{daily}}}$, 2.030 % to the main effects of $\overline{Tair_{1.3m_{daily}}}$, 3.528 % to the main effects of $\overline{G_{daily}}$ and 8.722, 4.846 and 6.017 % to the total effects of $\overline{Tair_{50m_{daily}}}$, $\overline{Tair_{1.3m_{daily}}}$ and $\overline{G_{daily}}$, respectively. LAI and thermal inertia also seem to be important parameters influencing the sensitivity of soil heat flux.

5 Discussion

In common with previous studies (e.g. Petropoulos et al. 2009c, 2010, 2013) this study has identified a small number of environmental parameters that appear to have a significantly impact on the SimSphere's outputs examined herein.

Table 3 Summarised SA results from the implementation of the BACCO GEM on the different outputs simulated by SimSphere

Model inputs		$\overline{Rg_{daily}}$		$\overline{Ldown_{daily}}$		$\overline{Lup_{daily}}$		$\overline{Tair_{50m_{daily}}}$		$\overline{Tair_{1.3m_{daily}}}$		$\overline{G_{daily}}$	
		Main effects (%)	Total effects (%)	Main effects (%)	Total effects (%)	Main effects (%)	Total effects (%)	Main effects (%)	Total effects (%)	Main effects (%)	Total effects (%)	Main effects (%)	Total effects (%)
X1	Slope	13.150	27.941	7.412	22.843	9.423	23.017	8.351	24.897	9.738	22.945	7.853	18.940
X2	Aspect	69.737	85.315	37.101	57.045	44.297	64.585	35.859	56.144	39.690	60.073	36.344	56.249
X3	Station height	0.003	0.003	0.045	0.087	0.105	0.374	0.061	0.142	0.025	0.381	0.018	0.019
X4	Fractional vegetation cover	0.478	0.693	14.910	21.743	0.208	2.281	15.137	21.318	0.033	0.116	10.611	17.274
X5	LAI	0.427	0.623	0.054	0.888	1.784	2.872	0.155	1.769	0.311	3.767	3.728	7.115
X6	Foliage emissivity	0.013	0.019	0.040	0.110	0.015	0.100	0.030	0.171	0.031	0.033	0.162	0.724
X7	[Ca]	0.002	0.002	0.038	0.447	0.029	0.030	0.027	0.371	0.027	0.821	0.101	0.551
X8	[Ci]	0.005	0.005	0.134	0.281	0.066	1.798	0.108	0.244	0.030	0.547	0.022	0.028
X9	[03] in the air	0.002	0.002	0.028	0.065	0.020	0.224	0.028	0.029	0.029	0.321	0.085	0.655
X10	Vegetation height	0.002	0.002	3.366	8.551	0.126	0.393	3.139	8.518	10.084	19.714	0.936	3.037
X11	Leaf width	0.003	0.002	0.060	0.219	0.017	0.018	0.055	0.472	0.067	0.111	0.105	0.570
X12	Minimum stomatal resistance	0.002	0.002	0.043	0.726	0.023	0.327	0.024	0.327	0.191	1.050	0.060	0.129
X13	Cuticle resistance	0.001	0.002	0.088	0.349	0.015	0.016	0.135	0.668	0.228	0.502	0.248	2.691
X14	Critical leaf water potential	0.003	0.015	0.061	0.595	0.035	0.042	0.096	1.059	0.019	0.114	0.036	0.101
X15	Critical solar parameter	0.004	0.005	0.036	0.281	0.097	0.190	0.069	0.226	0.101	0.888	0.090	0.361
X16	Stem resistance	0.003	0.040	0.014	0.015	0.029	0.030	0.014	0.015	0.080	0.117	0.056	0.124
X17	Surface moisture availability	0.002	0.002	6.515	8.466	16.049	21.823	6.760	8.722	2.030	4.846	3.528	6.017
X18	Root zone moisture availability	0.005	0.005	0.064	0.087	0.024	0.274	0.055	0.056	0.315	0.660	0.054	0.101
X19	Subst. max. volum. water cont.	0.003	0.003	0.029	0.099	0.101	0.155	0.034	0.035	0.111	0.215	0.067	0.355
X20	Subst. climatol. mean temp.	0.004	0.004	0.652	1.043	0.488	0.826	0.567	0.909	0.049	0.218	0.924	2.449
X21	Thermal inertia	0.001	0.002	0.034	0.035	0.038	0.284	0.029	0.030	0.019	0.020	5.896	13.708
X22	Ground emissivity	0.002	0.002	0.049	0.119	1.306	2.308	0.040	0.093	0.023	0.024	0.062	0.931
X23	Atmospheric precipitable water	0.002	0.003	0.037	1.114	0.120	0.762	0.042	0.632	0.017	0.215	0.061	0.691

(continued)

Table 3 (continued)

Model inputs		Rg_{daily}		$Ldown_{daily}$		Lup_{daily}		$Tair_{50m_{daily}}$		$Tair_{1.3m_{daily}}$		G_{daily}	
		Main effects (%)	Total effects (%)	Main effects (%)	Total effects (%)	Main effects (%)	Total effects (%)	Main effects (%)	Total effects (%)	Main effects (%)	Total effects (%)	Main effects (%)	Total effects (%)
X24	Surface roughness	0.003	0.003	2.018	6.478	0.106	0.796	1.599	6.091	6.937	18.416	0.602	1.548
X25	Obstacle height	0.004	0.005	0.024	0.025	0.021	0.473	0.035	0.036	0.038	0.170	0.083	0.084
X26	Fractional cloud cover	0.005	0.005	0.064	0.907	0.308	1.682	0.034	0.145	0.035	0.036	0.062	0.100
X27	RKS	0.059	0.801	0.017	0.018	0.038	0.086	0.018	0.019	0.072	0.331	0.025	0.026
X28	CosbyB	0.052	0.813	0.285	1.032	0.023	0.235	0.164	0.907	0.026	0.823	0.043	0.120
X29	THM	0.017	0.320	0.110	0.340	0.039	0.335	0.152	0.750	0.070	0.381	0.051	0.120
X30	PSI	0.022	0.508	0.025	0.026	0.373	4.824	0.018	0.019	0.028	0.498	0.064	0.229
Total % variance (main effects only)		84.016		73.351		75.285		72.385		70.452		71.974	
Total % variance contribution from first order interactions		14.160		11.642		10.528		11.997		16.171		19.031	
Total % variance (main effects only)		84.016		73.351		75.285		72.385		70.452		71.974	

Computed main and total effect indices by the GEM tool (expressed as %) for each of the model parameters are shown. The last three lines summarise the percentages of the explained total output variance of the main effects alone and after including the interaction effects. Italicized are highlights the model input parameters with greater than 1 % variance decomposition and main and total effects greater than 1

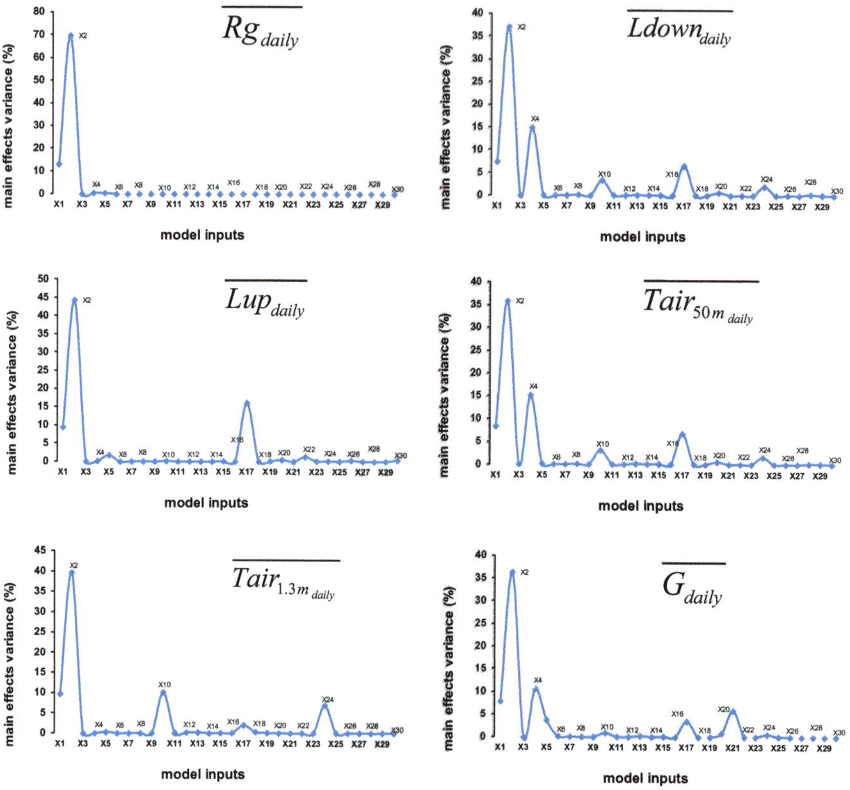

Fig. 3 An illustration of the decomposition of variance (main effects only) of the SimSphere model input parameters for the different target quantities examined

In particular, slope, aspect, fractional vegetation cover, and to a lesser extent surface soil moisture availability have been shown to be important. Several authors have noted the importance of slope and aspect in controlling $\overline{Rg_{daily}}$ (e.g. Oliphant et al. 2003; Petropoulos et al. 2009c, 2010, Sabetraftar et al. 2011). Both influence how much solar radiation is received at the Earth surface and the angle of incidence of that radiation. Similar reasoning also applies to $\overline{Ldown_{daily}}$. This primary influence of slope and aspect on incoming solar radiation will, through the net radiation balance, also influence $\overline{Lup_{daily}}$ emitted by the Earth's surface and $\overline{G_{daily}}$ as well as air temperatures through energy exchanges. The influence of fractional vegetation cover on these parameters is also to be expected and has also been found elsewhere (Petropoulos et al. 2013) because vegetation cover can influence how much incoming radiation (short wave or long wave) reaches the ground surface, and, depending on vegetation and soil albedo, the amount of radiation absorbed, reflected and emitted. Complexity is introduced where the influence of slope, aspect and vegetation cover interact and vary both spatially and temporally (Ellis and Pomeroy 2007; Ellis et al. 2011).

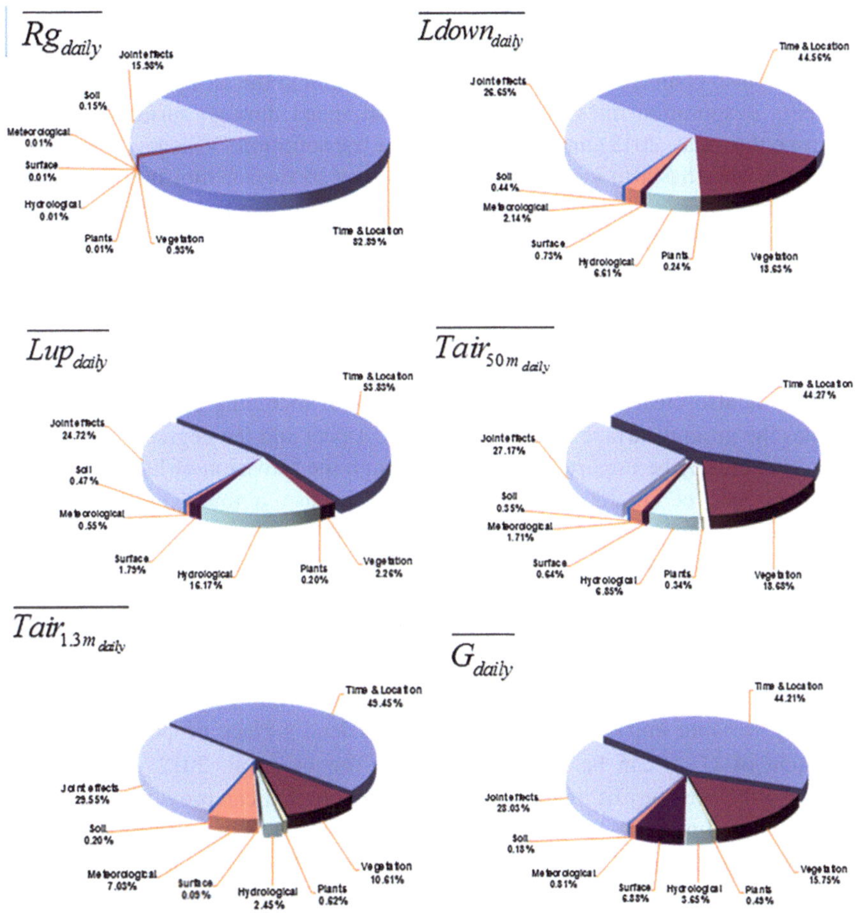

Fig. 4 An illustration of the decomposition of variance (main effects only) of the SimSphere model input parameters for the different groups of model input parameters (as those defined in the SimSphere menu)

It is also reasonable that air temperatures will be sensitive to the parameters to which incoming solar radiation is also sensitive (Fig. 3; Table 3), because they are dependent on radiation transfers through latent and sensible heat. Higher receipts of solar radiation result in higher sensible heat flux which will, in turn, lead to higher air temperatures (Durre et al. 2000). Since $\overline{Rg_{daily}}$ provides the source of energy for evaporation and models of surface temperature that are based on the fluxes of latent and sensible heat can be influences by changes in topography, especially land surface gradient (Sade et al. 2011). In addition, in the case of $\overline{Tair_{50m_{daily}}}$, the influence of fractional vegetation cover is to be expected since the proportion of vegetation cover may influence both latent heat and sensible heat flux and thus air temperature, as well as the amount of radiation reflected and emitted. In contrast the (possibly related) parameters of vegetation height and surface roughness seem to be more important influences on $\overline{Tair_{1.3m_{daily}}}$. Vegetation cover can also control the rate of transfer of radiation to the air by ground shading, reducing evaporation rates and reducing wind speed through surface roughness (Petropoulos et al. 2013) and increasing plant transpiration (Villegas et al. 2010). Other SA work has also identified the significance of surface moisture availability on air temperature (e.g. Carlson and Boland 1978; Petropoulos et al. 2009c, 2013) due in large part to its significant control on evapotranspiration (Santaneoll et al. 2009; Dirmeyer 2011; Lockart et al. 2012) and thus it's control on the partitioning of net radiation into latent and sensible heat. Limits on soil moisture limit evapotranspiration and lead the transfer of incoming solar radiation into sensible heat rather than latent heat.

$\overline{G_{daily}}$ is also clearly affected by the amount of incoming solar radiation that reaches the ground surface and, as such slope, aspect and the degree of vegetation cover will clearly be important, not least through the provision of litter on the soil surface (Ogee et al. 2001). Thermal inertia is also a much more important parameter for this output compared to other model outputs. Wang and Bou-Zeid (2012) note that soil moisture, through altering soil thermal properties, can influence thermal conductivity and heat capacity. It is therefore, to be expected that $\overline{G_{daily}}$ is sensitive to surface moisture availability. Numerous authors (e.g. Schwinger et al. 2010) have shown that $\overline{G_{daily}}$ is a significant element of energy balance fluxes and influences soil temperatures (van der Tol 2012). Due to the spatially variable impacts of soil moisture and land surface properties the measurement of $\overline{G_{daily}}$ can be very challenging (Tanguy et al. 2012) and as such, assessing the accuracy of models such as SimSphere that can provide such data as well as identifying the factors to which it is sensitive is an important research priority.

The applied significance of these results to other environmental fields is particularly evident in terms of the $\overline{Ldown_{daily}}$ because of its importance in the surface energy budget and, therefore climate change and global warming (Sicart et al. 2006). Measurements of this parameter is also key for cryospheric processes in the high latitudes and over glaciers due to its impact on snow and ice melt (Pomeroy et al. 2009; Bewley et al. 2010; Sicart et al. 2010; Juszak and Pelliccotti 2013). In terms of the importance of aspect input parameters, for example Sicart et al. (2006) noted that the sky view factor, influenced by aspect and slope, can influence long wave irradiance from the land surface and thus influence the energy balance. Pomeroy et al. (2009) showed that the influence of vegetation, in particular the vegetation canopy (i.e. fractional vegetation cover, leaf area index) strongly influence the surface energy budget, and particularly the contribution of short wave and long wave radiation. In their study on radiation fluxes over shrub cover in Canada Bewley et al. (2010) also showed that vegetation fraction and canopy transmissivity were particularly important in influencing the relative importance of short-wave and long-wave radiation on snow covered surfaces, but that crucially, complexity is introduced when snow cover modifies the vegetation cover.

Furthering our understanding and capability to forecast changes in land-surface-atmosphere interactions under a different climate regime in future requires that we develop models that can successfully integrate instrumented and remotely sensed data. This study has extended the sensitivity analysis of SimSphere to include six new output parameters and represents an important development in our understanding of the model. Once again, in common with the previous studies, the fact that a small number of environmental parameters influence the sensitivity of the model outputs suggests that if it is possible to obtain remotely-sensed and ground-truthed observations of these parameters at a regional scale then there is significant potential for the model. Some of these parameters (e.g. slope, aspect and fractional vegetation cover) can be obtained in a straightforward manner through EO methods, yet others (e.g. surface soil moisture availability) may present more challenges.

6 Conclusions

This study has contributed to further extending the sensitivity analysis of SimSphere SVAT model by analysing the sensitivity of six parameters for the first time. In common with previous studies, our work confirms that only a small number of input parameters exert a strong control on model outputs. Slope and aspect are clearly very important input parameters for the simulation of the model outputs considered in this stud due to their influence on incoming solar radiation and resultant energy fluxes and hydrological processes at or near the Earth's surface. Vegetation height, surface moisture availability and surface roughness are also important input parameters. In the context of developing SimSphere itself as a stand-alone model as well as a tool which can be integrated with remotely sensed data, this study has further underlined the fact that if accurate, remotely sensed measurements of a small number of key input parameters can be obtained, its potential for contributing to our understanding of regional-scale land-surface interactions is significant. Further work should be directed towards performing an all-inclusive SA of all SimSphere parameters as well as comparing the model outputs against corresponding in situ data across different ecosystems globally.

Acknowledgments Preparation of this work was conducted under the Marie Curie Career Re-Integration Project TRANSFORM-EO project. Dr. Petropoulos gratefully acknowledges the financial support provided.

References

Battrick B (2006) The changing Earth. New scientific challenges for ESA's living planet programme. ESA SP-1304. ESA Publications Division, ESTEC, The Netherlands

Bewley D, Essery R, Pomeroy J, Ménard C (2010) Measurements and modelling of snowmelt and turbulent heat fluxes over shrub tundra. Hydrol Earth Syst Sci 14(7):1331–1340

Carlson TN (2007) An overview of the "triangle method" for estimating surface evapotranspiration and soil moisture from satellite imagery. Sensors 7:1612–1629

Carlson TN, Boland FE (1978) Analysis of urban-rural canopy using a surface heat flux/temperature model. J Appl Meteorol 17:998–1014

Chauhan NS, Miller S, Ardanuy P (2003) Spaceborne soil moisture estimation at high resolution: a microwave-optical/IR synergistic approach. Int J Remote Sens 22:4599–4646

Chen L, Tian Y, Cao C, Zhang S, Zhang S (2012) Sensitivity and uncertainty analyses of an extended ASM3-SMP model describing membrane bioreactor operation. J Membr Sci 389:99–109

Chen Y, Yu J, Khan S (2010) Spatial sensitivity analysis of multi-criteria weights in GIS-based land suitability evaluation. Environ Model Softw 25:1582–1591

Dirmeyer PA (2011) The terrestrial segment of soil moisture–climate coupling. Geophys Res Lett 38(16):L16702

Durre I, Wallace JM, Lettenmaier DP (2000) Dependence of extreme daily maximum temperatures on antecedent soil moisture in the contiguous United States during summer. J Clim 13:2641–2651

Ellis CR, Pomeroy JW (2007) Estimating sub-canopy shortwave irradiance to melting snow on forested slopes. Hydrol Process 21(19):2581–2593

Ellis CR, Pomeroy JW, Essery RLH, Link TE (2011) Effects of needleleaf forest cover on radiation and snowmelt dynamics in the Canadian Rocky Mountains. Can J For Res 41:608–620

Feyissa AH, Gernaey KV, Adler-Nissen J (2012) Uncertainty and sensitivity analysis: mathematical model of coupled heat and mass transfer for a contact baking process. J Food Eng 109:281–290

Gatelli D, Kucherenko S, Ratto M, Tarantola S (2009) Calculating first-order sensitivity measures: a benchmark of some recent methodologies. Reliab Eng Syst Saf 94:1212–1219

Gillies RR, Carlson TN, Cui J, Kustas WP, Humes KS (1997) Verification of the triangle method for obtaining surface soil moisture content and energy fluxes from remote measurements of the normalised difference vegetation index (NDVI) and surface radiant temperatures. Int J Remote Sens 18:3145-3166

Holvoet K, van Griensven A, Seuntjents P, Vanrolleghem PA (2005) Sensitivity analysis for hydrology and pesticide supply towards the river in SWAT. Phys Chem Earth 30:518–526

Johnson JS, Gosling JP, Kennedy MC (2011) Gaussian process emulation for second-order Monte Carlo simulations. J Stat Plann Infer 141:1838–1848

Juszak I, Pellicciotti F (2013) A comparison of parameterisations of incoming longwave radiation over melting glaciers: model robustness and seasonal variability. J Geophys Res Atmos 118:3066–3084. http://dx.doi.org/10.1002/jgrd.50277

Kennedy MC, Butler Ellis MC, Miller PCH (2012) BREAM: a probabilistic Bystander and Resident Exposure Assessment Model of spray drift from an agricultural boom sprayer. Comput Electron Agric 88:63–71

Kennedy MC, O'Hagan A (2000) Predicting the output from a complex computer code when fast approximations are available. Biometrika 87:1–13

Kennedy MC, O'Hagan A (2001) Bayesian calibration of computer models. J Roy Stat Soc: Ser. B (Stat Methodol) 63(3):425–464

Lockart N, Kavetski D, Franks SW (2012) On the role of soil moisture in daytime evolution of temperatures. Hydrol Process doi:10.1002/hyp.9525

Moradkhani H (2008) Hydrologic remote sensing and land surface data assimilation. Sensors 8(5):2986–3004

Moran MS, Peters-Lidard CD, Watts JM, McElroy S (2004) Estimating soil moisture at the watershed scale with satellite-based radar and land surface models. Can J Remote Sens 30(5):805–824

Nossent J, Elsen P, Bauwens W (2011) Sobol's sensitivity analysis of a complex environmental model. Environ Model Softw 26:1515–1525

O'Hagan A (2006) Bayesian analysis of computer code outputs: a tutorial. Reliab Eng Syst Saf 91:1290–1300
Ogée, Lamaud J, Brunet E, Berbigier Y, Bonnefond P (2001) A long-term study of soil heat flux under a forest canopy. Agric For Meteorol 106(3):173–186
Olioso A (1992) Simulation des 6changes d'6nergie et de masse d'un convert v6gandal, dans le but de relier ia transpiration et la photosynthese anx mesures de reflectance et de temp6rature de surface. Thesis, University de Montpellier II
Olioso A, Chauki H, Courault D, Wigneron J-P (1999) Estimation of evapotranspiration and photosynthesis by assimilation of remote sensing data into SVAT models. Remote Sens Environ 68:341–356
Oliphant AJ, Spronken-Smith RA, Sturman AP, Owens IF (2003) Spatial variability of surface radiation fluxes in mountainous terrain. J Appl Meteor 42:113–128
Parry HR, Topping CJ, Kennedy MC, Boatman ND, Murray AWA (2012) Bayesian sensitivity analysis applied to an agent-based model of bird population response to landscape change. Environ Model Softw 45(6):1–12
Petropoulos G, Carlson, TN (2011) Retrievals of turbulent heat fluxes and soil moisture content by Remote Sensing. In: Weng Q. (ed) Advances in environmental remote sensing: sensors, algorithms, and applications, Taylor and Francis, pp 667–502 ISBN: 978-1-4200-9175-5
Petropoulos GP (2013) Remote sensing of energy fluxes and soil moisture content, Taylor and Francis, 506 pp. ISBN: 978-1-4665-0578-0
Petropoulos GP, Carlson TN (2013) Deriving surface soil moisture from medium resolution VNIR/TIR Earth observation data and a 1D process model. European Geosciences Union, Vienna, Austria
Petropoulos G, Carlson TN, Wooster MJ, Islam S (2009a) A review of Ts/VI remote sensing based methods for the retrieval of land surface fluxes and soil surface moisture content. Adv Phys Geogr 33(2):1–27
Petropoulos G, Carlson TN, Wooster MJ (2009b) An overview of the use of the SimSphere soil vegetation atmosphere transfer (SVAT) model for the study of land-atmosphere interactions. Sensors 9(6):4286–4308
Petropoulos G, Wooster MJ, Kennedy M, Carlson TN, Scholze M (2009c) A global sensitivity analysis study of the 1D SimSphere SVAT model using the GEM SA software. Ecol Model 220(19):2427–2440
Petropoulos G, Ratto M, Tarantola S (2010) A comparative analysis of emulators for the sensitivity analysis of a land surface process model. Procedia—Soc Behav Sci 2(6):7716–7717 (6th international conference on sensitivity analysis of model output, 19–22 July 2010, Milan, Italy). doi:10.1016/j.sbspro.2010.05.194
Petropoulos GP, Griffiths HM, Tarantola S (2013) Towards operational products development from Earth observation: exploration of SimSphere land surface process model sensitivity using a GSA approach. In: 7th International conference on sensitivity analysis of model output, Nice, France, 1–4 July 2013 (accepted)
Piles M, Camps A, Vall-llossera M, Corbella I, Panciera R, Rudiger C, Kerr YH, Walker J (2011) Downscaling SMOS-derived soil moisture using MODIS visible/infrared data. IEEE Trans Geosci Remote Sens 49(9):3156–3166
Pomeroy JW, Marks D, Link T, Ellis C, Hardy J, Rowlands A, Granger R (2009) The impact of coniferous forest temperature on incoming longwave radiation to melting snow. Hydrol Process 23(17):2513–2525
Ratto M, Tarantola S, Saltelli A (2011) Sensitivity analysis in model calibration: GSA-GLUE approach. Comput Phys Commun 136:212–224
Sabetraftar K, Mackey B, Croke B (2011) Sensitivity of modelled gross primary productivity to topographic effects on surface radiation: a case study in the Cotter River Catchment, Australia. Ecol Model 222(3):795–803
Sade R, Rimmer A, Iggy Litaor M, Shamir E, Furman A (2011) The sensitivity of snow-surface temperature equation to sloped terrain. J Hydrol 408(3):308–313

Saltelli A, Chan K, Scott EM (2000) Sensitivity analysis. In: Wiley series in probability and statistics. Wiley, Chichester, P 467
Saltelli A, Tarantola S, Chan KP-S (1999) A quantitative model-independent method for global sensitivity analysis of model output. Technometrics 41(1):39–56
Santanello JA, Peters-Lidard CD, Kumar SV, Alonge C, Tao W-K (2009) A modeling and observational framework for diagnosing local land–atmosphere coupling on diurnal time scales. J Hydrometeorol 10:577–599
Schwieger V (2004) Variance-based sensitivity analysis for model evaluation in engineering surveys. INGEO 2004 and FIG Regional central and Eastern European conference on engineering surveying, Bratislava, Slovakia, 11–13 Nov 2004. Available on line at: http://www.fig.net/pub/bratislava/papers/ts01/ts01schwieger.pdf. Accessed 24 April 2007
Schwinger J, Kollet SJ, Hoppe CM, Elbern H (2010) Sensitivity of latent heat fluxes to initial values and parameters of a land-surface model. Vadose Zone J 9(4):984–1001
Sicart JE, Pomeroy JW, Essery RLH, Bewley D (2006) Incoming longwave radiation to melting snow: observations, sensitivity and estimation in northern environments. Hydrol Process 20(17):3697–3708
Sicart JE, Hock R, Ribstein P, Chazarin JP (2010) Sky longwave radiation on tropical Andean glaciers: parameterization and sensitivity to atmospheric variables. J Glaciol 56(199):854–860
Sobol' IM (1976) Uniformly distributed sequences with additional uniformity properties. Zh Vychisl Mat Mat Fiz 16(5):1332–1337
Song X, Bryan BA, Paul KI, Zhao G (2012) Variance-based sensitivity analysis of a forest growth model. Ecol Model 246:135–143
Tanguy M, Baille A, González-Real MM, Lloyd C, Cappelaere B, Kergoat L, Cohard JM (2012) A new parameterisation scheme of ground heat flux for land surface flux retrieval from remote sensing information. J Hydrol 454:113–122
van der Tol C (2012) Validation of remote sensing of bare soil ground heat flux. Remote Sens Environ 121:275–286
Villegas JC, Breshears DD, Zou CB, Law DJ (2010) Ecohydrological controls of soil evaporation in deciduous drylands: how the hierarchical effects of litter, patch and vegetation mosaic cover interact with phenology and season. J Arid Environ 74(5):595–602
Wang ZH, Bou-Zeid E (2012) A novel approach for the estimation of soil ground heat flux. Agric For Meteorol 154:214–221

Remote Estimation of Land Surface Temperature for Different LULC Features of a Moist Deciduous Tropical Forest Region

Suman Sinha, Prem Chandra Pandey, Laxmi Kant Sharma, Mahendra Singh Nathawat, Pavan Kumar and Shruti Kanga

Abstract Potential of Landsat TM thermal sensor is investigated to retrieve land surface temperature (LST) using spectral index (NDVI), spectral radiance and surface emissivity for a moist deciduous tropical forest area of Munger forests (Bihar, India). Surface emissivity values derived from NDVI are directly used for LST estimation. LST varies spatially due to the complexity of land surface cover features and helps in land-use/land-cover profiling. Areas covered with vegetation show minimum temperatures; while barren and exposed land shows high values. Built-up land generally has higher LST, but when dispersed in small pockets in the forests, the LST value decreases as revealed in the results.

Keywords Landsat · Forest surface temperature (FST) · Normalized difference vegetation index (NDVI) · Land-use/land-cover (LULC) · Emissivity

---

S. Sinha
Department of Remote Sensing, Birla Institute of Technology,
Ranchi, Jharkhand 835215, India
e-mail: sumanrumpa.sinha@gmail.com

P. C. Pandey (✉)
Centre for Landscape and Climate Research, Department of Geography,
University of Leicester, Leicester LE1 7RH, UK
e-mail: prem26bit@gmail.com

L. K. Sharma · S. Kanga
Centre for Land Resource Management, Central University of Jharkhand,
Ranchi, Jharkhand 835215, India
e-mail: laxmikant1000@yahoo.com

M. S. Nathawat
School of Sciences, Indira Gandhi National Open University (IGNOU),
Maidan Garhi, New Delhi 110068, India
e-mail: msnathawat@ignou.ac.in

P. Kumar
Department of Remote Sensing, Banasthali University, Tonk, Rajasthan 304022, India
e-mail: pawan2607@gmx.com

P. K. Srivastava et al. (eds.), *Remote Sensing Applications in Environmental Research*,
Society of Earth Scientists Series, DOI: 10.1007/978-3-319-05906-8_4,

1 Introduction

Land surface temperature (LST) is controlled by several physical and chemical thermodynamics, surface and atmospheric parameters and on the other hand, LST controls several physical, chemical and biological processes of the Earth (Becker and Li 1990). It can provide important information about the surface physical properties and climate which plays a role in many environmental and meteorological processes (Dousset and Gourmelon 2003; Weng et al. 2004). Moreover, changes in land-use/land-cover (LU/LC) and anthropogenic activities affect the climate. Conventional methods like measuring air temperature with land observing stations and temperature sensors mounted on vehicles for estimating temperatures are expensive and time taking and can even lead to problems of spatial interpolation. Remote sensing becomes a better alternative in this context. Possibly the best way to retrieve this parameter regionally and globally is the use of satellite remote sensing due to the availability of high resolution, consistent and repetitive coverage and capability of measurements of earth surface conditions (Owen et al. 1998). Dash et al. (2001, 2002) gave a detailed conceptualization of the physics and theory behind deriving LST. Quantitative estimation of surface temperature along different LULC features can be made by using thermal infrared (TIR) sensors mounted on remote sensing satellites. LST is sensitive to vegetation and soil moisture; and hence can detect changes in LULC features (Mallick et al. 2008).

Valuable studies related to Urban Heat Island effect are carried out using integrated geospatial techniques (Sarkar 2004; Weng et al. 2004; Fan et al. 2007; Stathopoulou and Cartalis 2007; Jenerette et al. 2007; Li and Yu 2008; Hais and Kučera 2008; Cheval and Dumitrescu 2009). MODIS is used widely for estimation of LST for small-scale global scenario (Galve et al. 2007; Pinheiro et al. 2007; Akhoondzadeh and Saradjian 2008; Hanes and Schwartz 2011; van Leeuwen et al. 2011; Mildrexler et al. 2011; Hachem et al. 2012). The use of Landsat Thematic Mapper (TM) and ETM+ thermal data is better for large-scale regional and local set-up. Thermal infrared band (10.44–12.42 μm) present in TM/ETM+ with high spatial resolution (120 m for TM and 60 m for ETM+) are much useful for local and regional thermal infrared study. Several researches targeted in the retrieval of LST from Landsat 4 and 5 thermal infrared data (Schott and Volchok 1985; Wukelic et al. 1989; Goetz et al. 1995). Innovative robust however easy techniques are required to attain high accuracy in prediction of LST from TM/ETM+ TIR data with fewer parameters. Earth's surface encompasses various complicated land-use and land-cover feature types. It is difficult to accurately determine the surface emissivities of these features. An error of approximately 6 °C could occur for an assumed emissivity of 1; while, maximum error of only 0.8 °C can be attained with adjustments made on emissivity values (Sutherland and Bartholic 1979). The incoming solar radiation at the Earth's surface results due to the complex interaction of energy between the Earth's surface and atmosphere is also an important parameter (Dubayah and Rich 1995). Moreover, spectral vegetation indices have great potential to be used as an alternative for the estimation of surface temperature (Faris and Reddy 2010). Several studies found certain existence of relationship between LST and Normalized Difference Vegetation Index (NDVI) (Kaufmann et al. 2003; Sun and Kafatos 2007; Yue et al. 2007). Light Use Efficiency (LUE) model has been established using NDVI, EVI2 and SAVI, in association with scaled LST (Wu and Niu 2012).

New climate policies and agreements like REDD (Reducing Emissions from Deforestation and Degradation) demands for monitoring the forests accurately (van Leeuwen et al. 2011). A detailed review of the role of remote sensing in estimation of LST for meteorological and climatological studies, including the different thermal remote sensing sensors to measure LST is documented by Tomlinson et al. (2011). The present studies investigate the potential of Landsat 5 TM in estimating LST over heterogeneous moist tropical forest area and make LULC profiling based on the information retrieved. Spatial analysis was carried out by building models involving vegetation spectral indices like NDVI with surface emissivity to estimate the surface temperature. This can be used to determine the canopy surface temperature or the forest surface temperature (FST), which is unfeasible with conventional methods.

2 Materials and Methods

2.1 Study Area and Datasets

A heterogeneous tropical moist deciduous forest of Munger situated in Bihar (India) is selected as a case study (Fig. 1). Satellite data of Landsat 5 TM was downloaded from USGS (http://www.usgs.gov/). Along with the satellite imagery, Survey of India (SOI) toposheets were also used for classifying LULC features.

2.2 Image Interpretation for LULC

The Landsat TM image is co-registered and geometrically corrected in reference to already registered Landsat ETM+ image of the same region. The image was even corrected with certain distinct identifiable objects on the ground. Features on ground are correlated to image features based on spectral information which is a standard visual technique for interpretation based on image elements, the information of which are collected from ground. Training sets are selected from the FCC imagery for supervised classification with Maximum Likelihood Classification (MLC) based on the collected sample points for respective LULC classes. Figure 2 shows the LULC classification of the study area.

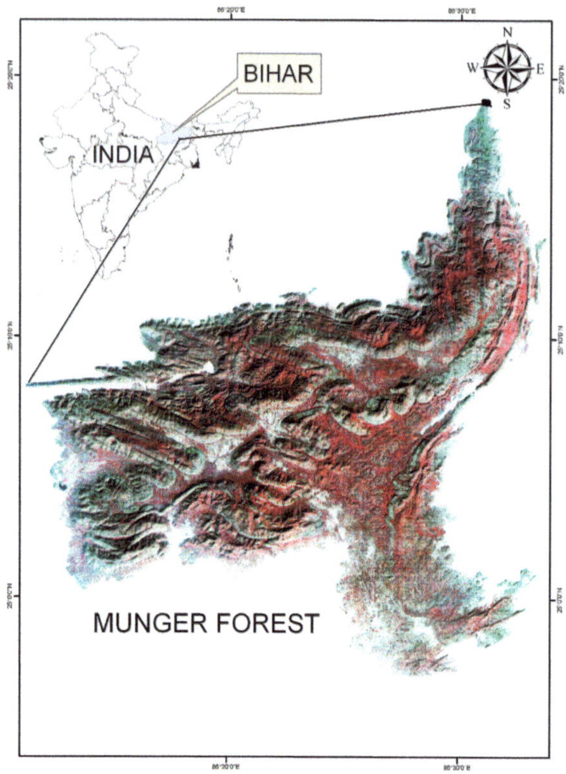

Fig. 1 Location of the study area of Munger forests in Bihar (India)

2.3 Surface Temperature Estimation

Planck's law states the dependence of spectral radiance (L_λ) at a certain spectral band with wavelength λ emitted from a blackbody [i.e., surface emissivity ε $(\lambda) = 1$] on the body's kinetic temperature and the determination of LST is based on this (Mildrexler et al. 2011). At-sensor black body brightness temperature (T_b) is calculated (Weng et al. 2004; Fan et al. 2007; NASA 2008). The calibration for TIR band is executed in a two-step process (Landsat Project Science Office 2002). First step, as shown in Eq. 1, comprises the conversion of TIR digital number (DN) values into L_λ ($Wm^{-2}\ sr^{-1}\ \mu m^{-1}$) (NASA 2009). All the equation variables are provided in the metadata file when the image is downloaded from the website.

$$L_\lambda = \left(\frac{L_{max} - L_{min}}{QCAL_{max} - QCAL_{min}}\right) * (DN - QCAL_{min}) + L_{min} \tag{1}$$

where,

DN pixel value

L_{min} spectral radiance scaled to $QCAL_{min}$ in $Wm^{-2}\ sr^{-1}\ \mu m^{-1}$

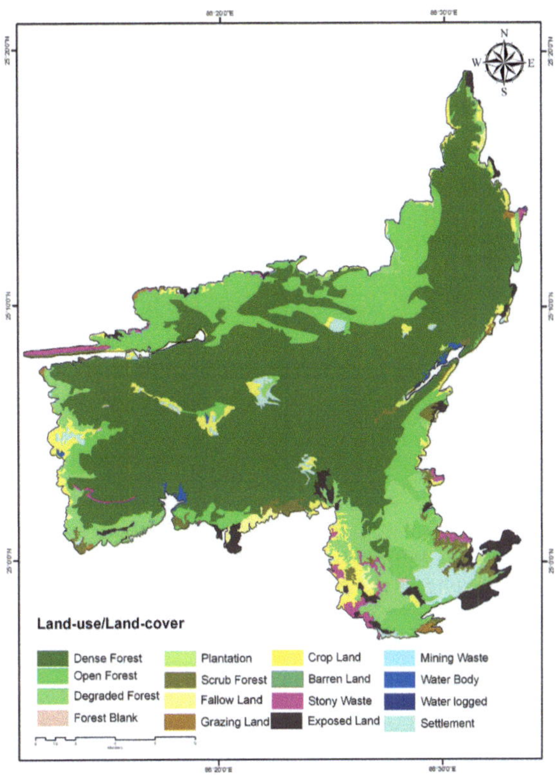

Fig. 2 Land-use/land-cover classification map of the study area classified into 16 distinct classes

L_{max} spectral radiance scaled to $QCAL_{max}$ in $Wm^{-2}\ sr^{-1}\ \mu m^{-1}$

$QCAL_{min}$ minimum quantized calibrated pixel value (corresponding to L_{min}) in DN

$QCAL_{max}$ maximum quantized calibrated pixel value (corresponding to L_{max}) in DN.

Thereafter, this L_λ is converted to T_b in Kelvin using Eq. 2 (Weng et al. 2004; Stathopoulou and Cartalis 2007; Fan et al. 2007; NASA 2008; Faris and Reddy 2010).

$$T_b = \frac{K_2}{\ln\left[\left(\frac{K_1}{L_\lambda}\right) + 1\right]} \tag{2}$$

where, T_b is the at-sensor brightness temperature in K; K_1 and K_2 are the calibration constants having values 607.76 $Wm^{-2}\ sr^{-1}\ \mu m^{-1}$ and 1,260.56 K respectively.

The at-sensor brightness temperature attained by the blackbody generates same radiance at the same wavelength of $\lambda = 11.5\ \mu m$. So, estimation of spectral emissivity requires further corrections to target the non-uniform emissivity depending upon the nature of the land surface. Surface emissivity of an object depicts how effectively the object is able to radiate energy in comparison to a black body (Lillesand and Kiefer 2003). This is controlled by physicochemical parameters and for vegetative areas floral species composition and phenology can exert significant impacts on the emissivity (Snyder et al. 1998; Yue et al. 2007). Surface emissivity is a function of spectral emissivity (Weng et al. 2004). NDVI can be used for emissivity correction (Eq. 3) for the different land covers estimated (Oberg and Assefa 2006; Duah et al. 2008; Faris and Reddy 2010).

$$\varepsilon = 0.047 \times \ln(NDVI) + 1.009 \tag{3}$$

The Normalized Difference Vegetation Index (NDVI) is calculated from visible and near infrared bands of TM 5 imagery as follows (Lillesand and Kiefer 2003):

$$NDVI = \frac{(NIR - R)}{(NIR + R)} \tag{4}$$

Finally, the emissivity corrected land surface temperature (T_s) is calculated using Eq. 5 (Weng et al. 2004; Weng and Yang 2006; Faris and Reddy 2010):

$$T_s = \frac{T_b}{1 + (\lambda \times T_b/\gamma) \times \ln \varepsilon} \tag{5}$$

where

T_s is the land surface temperature (K);

T_b is at-sensor brightness temperature (K);

ε is the spectral surface emissivity;

λ is the average of limiting wavelength of emitted radiance of ETM+ band 6 (11.5 μm);

$$\gamma = h \times c/\alpha\,(0.01438\,\mathrm{m\,K})$$

where,

α Boltsmann constant (1.38×10^{-23} J K)

h Plank's constant (6.626×10^{-34} J s)

c velocity of light (2.998×10^{8} m/s).

Atmospheric interference also hampers the accurate estimation of surface temperature; however, the error in the emissivity correction is two times larger than the error in the atmospheric correction (Prata et al. 1995). We are concerned in the relative surface temperature differences between different LULC features, hence, the error due to the atmospheric interferences is not considered.

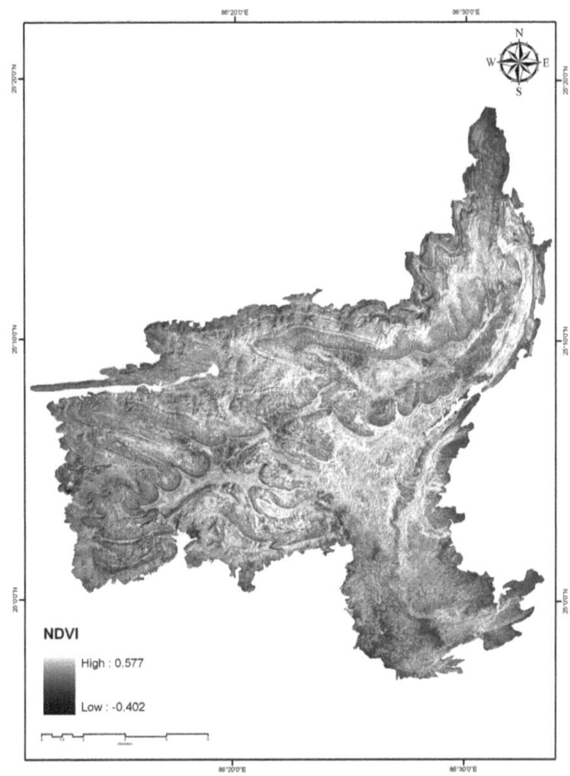

Fig. 3 Normalized difference vegetation index (NDVI) map

3 Results and Discussions

Landsat 5 TM satellite imagery is used to compute various spatial parameters like NDVI, Surface Emissivity, Land Surface Temperature and LULC map of Munger. LULC classification map (Fig. 2) is generated in ArcGIS based on field knowledge and reconnaissance survey. The satellite data and SOI toposheets are also used as references for proper LULC feature identifications. NDVI, as calculated using Eq. 4 ranges from −0.4 to 0.57 (Fig. 3) and shows negative values for water bodies, while positive for vegetative regions; the values increase on increase of vegetative portions. Figure 4 shows surface emissivity of LULC features which is high for vegetative regions and low for non-vegetative barren and exposed lands and water bodies. The spatial distribution of surface temperature in Munger (Fig. 5) ranges from 23 to 24 °C in the dense vegetation areas to a maximum of 28–30 °C at fallow, grazing and barren open spaces and rocky wastes; however, maximum for the exposed areas. Hence, the temperature difference among different LULC classes reaches nearly 6–7 °C. Maximum proportion of the forest is dense with small patches of heterogeneously distributed settlements along with scattered patches of agricultural fields. The existence of abundant vegetation in the

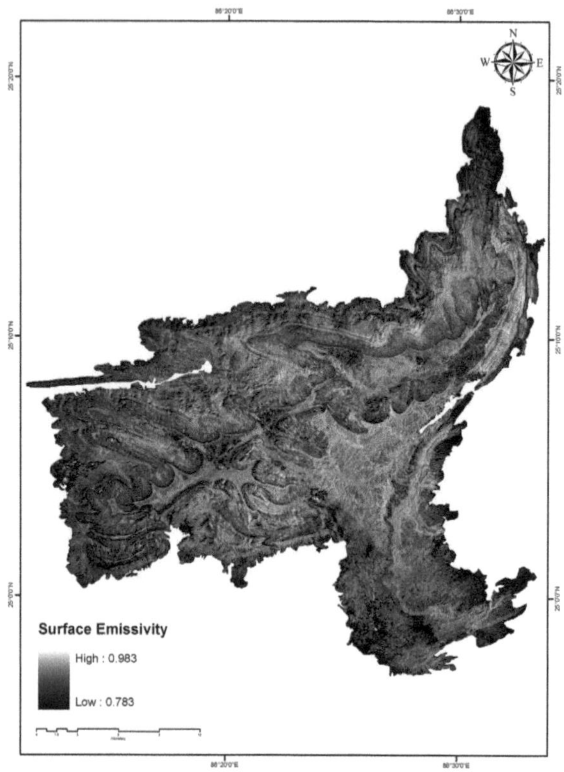

Fig. 4 Surface emissivity modeled in ArcGIS

area reduces the radiation heat flux of the earth surface by consuming a part of the radiation energy during evapotranspiration. However, forest blanks and gaps show higher values as these small man-made patches are partial scrub lands meant for grazing domestic wildlife. ST for water bodies and water logged areas are to some extent high as compared to the vegetation areas due to high surface emissivity. Settlements within the study area are scattered in heterogeneous pockets and hence, the vegetation cover reduces the surface temperature. Plantation, scrub land, stony waste and agricultural (crop lands) areas have moderate values, as obvious. Plantations are not dense; hence reflecting more emissivity, while the values of ST for crop lands depends on the phenology of the crop and the growing season. Scrub forests have less canopy density with areas covered with herbs, shrubs and bushes. Stony wastes show an average ST value of more than 28 °C, which is greater than plantations (27.78 °C), shrub forest (27.7 °C) and crop lands (27.52 °C). Mining waste on the other hand shows very low average ST values of nearly 25 °C, probably due to greater absorption of heat radiations. Figure 6 represents the graph showing the mean values of ST for different LULC categories of the area that can serve for LULC profiling.

Factors affecting LST retrieval from satellite TIR data like, radiance, transmittance, emissivity, atmospheric moisture, etc. are not that easy to estimate. We

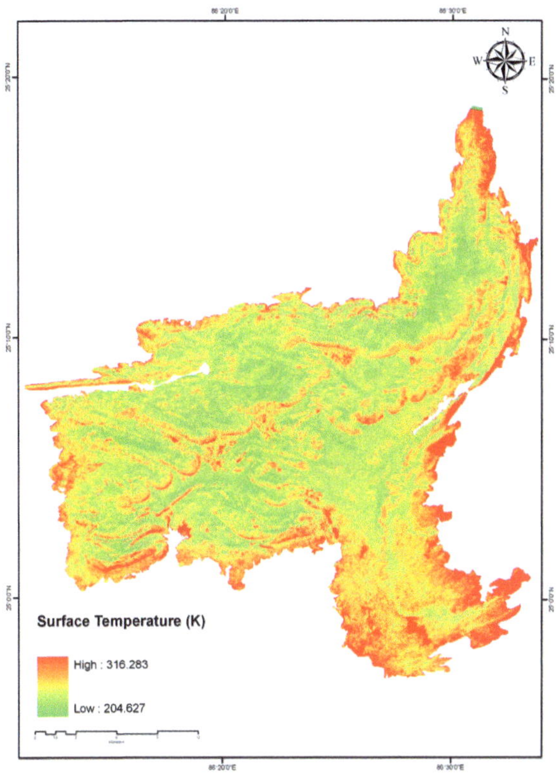

Fig. 5 Land surface temperature (LST) in Kelvin

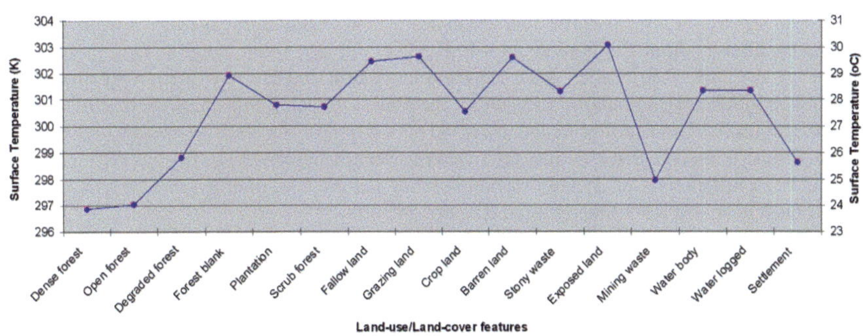

Fig. 6 LULC profiling based on ST derived through geospatial techniques. The graph shows average ST values that are represented both in Kelvin (K) and degree Celcius (°C) for each LULC categories reflecting their trend

have selected surface emissivity as not only an important factor; this can also be easily assessed through remote sensing. Moreover, emissivity is characterized in terms of NDVI that is determined through remote sensing techniques. NDVI vary spatially with the spatial features on ground. Retrieval of surface temperature of heterogeneous tropical forests by Landsat TM shows relative low ranges of ST in the forested vegetative parts, which is quite obvious. Water surfaces cause reduction of radiation heat flux and hence the water bodies and water-logged areas show lower ST values as compared to exposed lands and barren areas; however, higher than forests. Low ST values of rural settlements accounts due to the fact that they are distributed in small scattered pockets within and around the forested regions. Lack of economic developments in terms of concrete roads and structures restrict the temperature values. The values are high in fallow lands, barren spaces and exposed areas. Along with the proper estimation of forest canopy surface temperature (FST), Landsat TIR data can simultaneously be used to retrieve temperatures within a forest, which is unfeasible with conventional or direct field-based methods. The method can be even upscaled for applications in larger areas. Hence, the method adopted is a simple approach to predict LST with considerable accuracy involving fewer parameters.

Acknowledgments The authors express sincere gratitude to the DST, Government of India for providing funds to carry out the research, Officials of Project Tiger, Sariska and Sariska Forest Division are acknowledged for their support.

References

Akhoondzadeh M, Saradjian MR (2008) Comparison of land surface temperature mapping using MODIS and ASTER images in semi-arid area. Int Arch Photogram Remote Sens Spat Inf Sci XXXVII Part B8:873–876

Becker F, Li Z-L (1990) Temperature independent spectral indices in thermal infrared bands. Remote Sens Environ 32:17–33

Cheval S, Dumitrescu A (2009) The July urban heat island of Bucharest as derived from MODIS images. Theoret Appl Climatol 96:145–153

Dash P et al (2001) Retrieval of land surface temperature and emissivity from satellite data: physics, theoretical limitations and current methods. J Indian Soc Remote Sens 29:23–30

Dash P et al (2002) Land surface temperature and emissivity estimation from passive sensor data: theory and practice-current trends. Int J Remote Sens 23:2563–2594

Dousset B, Gourmelon F (2003) Satellite multi-sensor data analysis of urban surface temperatures and landcover. ISPRS J Photogramm Remote Sens 58:43–54

Duah SO, Donoghue DNM, Burt TP (2008) Intercomparison of evapotranspiration over the Savannah Voltu Basin in West Africa using remote sensing data. Sensors 8:2736–2761

Dubayah R, Rich PM (1995) Topographic solar radiation models for GIS. Int J Geogr Inf Syst 9:401–419

Fan L et al (2007) Regional land surface energy fluxes by satellite remote sensing in the Upper Xilin River Watershed (Inner Mongolia, China). Theoret Appl Climatol 88:231–245

Faris AA, Reddy YS (2010) Estimation of urban heat island using Landsat ETM+ imagery at Chennai city—a case study. Int J Earth Sci Eng 3:332–340

Galve JM et al (2007) Simulation and validation of land surface temperature algorithms for MODIS and AATSR data. Tethys 4:27–32

Goetz SJ et al (1995) Surface temperature retrieval in a temperate grassland with multiresolution sensors. J Geophys Res 100:25397–25410

Hachem S, Duguay CR, Allard M (2012) Comparison of MODIS-derived land surface temperatures with ground surface and air temperature measurements in continuous permafrost terrain. Cryosphere 6:51–69
Hais M, Kučera T (2008) Surface temperature change of spruce forest as a result of bark beetle attack: remote sensing and GIS approach. Eur J For Res 127:327–336
Hanes JM, Schwartz MD (2011) Modeling land surface phenology in a mixed temperate forest using MODIS measurements of leaf area index and land surface temperature. Theoret Appl Climatol 105:37–50
Jenerette GD et al (2007) Regional relationships between surface temperature, vegetation, and human settlement in a rapidly urbanizing ecosystem. Landscape Ecol 22:353–365
Kaufmann RK et al (2003) The effect of vegetation on surface temperature: a statistical analysis of NDVI and climate data. Geophys Res Lett 30:2147. doi:10.1029/2003GL018251
Landsat Project Science Office (2002) Landsat 7 science data user's handbook. Goddard Space Flight Center. Available from http://ltpwww.gsfc.nasa.gov/IAS/handbook/handbook_toc.html
Li K, Yu Z (2008) Comparative and combinative study of Urban Heat island in Wuhan city with remote sensing and CFD simulation. Sensors 8:6692–6703
Lillesand TM, Kiefer RW (2003) Remote sensing and image interpretation, 4th edn. Wiley, New York
Mallick J, Kant Y, Bharath BD (2008) Estimation of land surface temperature over Delhi using Landsat-7 ETM+. J Indian Geophys Union 12:131–140
Mildrexler DJ, Zhao M, Running SW (2011) A global comparison between station air temperatures and MODIS land surface temperatures reveals the cooling role of forests. J Geophys Res 116:G03025. doi:10.1029/2010JG001486
NASA (2008) Landsat-7 science data users handbook, Chap. 11. Available from http://Landsathandbook.gsfc.nasa.gov/handbook.html
NASA (2009) Landsat 7 science data users handbook, Chap. 2. 11:117–120. Available from http://Landsathandbook.gsfc.nasa.gov/handbook.html
Oberg JW, Assefa AM (2006) Evapotranspiration dynamics at an ecohydrological restoration site: an energy balance and remote sensing approach. J Am Water Resour Assoc 42:565–582
Owen TW, Carlson TN, Gillies RR (1998) Remotely sensed surface parameters governing urban climate change. Int J Remote Sens 19:1663–1681
Pinheiro ACT et al (2007) Near-real time retrievals of land surface temperature within the MODIS rapid response system. Remote Sens Environ 106:326–336
Prata AJ et al (1995) Thermal remote sensing of land surface temperature from satellites: current status and future prospects. Remote Sens Rev 12:175–224
Sarkar H (2004) Study of land cover and population density influences on urban heat island in tropical cities by using remote sensing and GIS: a methodological consideration. Paper presented at the 3rd FIG Regional Conference Jakarta, Indonesia, 3–7 October 2004
Schott JR, Volchok WJ (1985) Thematic mapper thermal infrared calibration. Photogram Eng Remote Sens 51:1351–1357
Snyder WC et al (1998) Classification based emissivity for land surface temperature measurement from space. Int J Remote Sens 19:2753–2774
Stathopoulou M, Cartalis C (2007) Daytime urban heat islands from Landsat ETM+ and Corine land cover data: an application to major cities in Greece. Sol Energy 81:358–368
Sun D, Kafatos M (2007) Note on the NDVI-LST relationship and the use of temperature-related drought indices over North America. Geophys Res Lett 34:L24406. doi:10.1029/2007GL031485
Sutherland RA, Bartholic JF (1979) Emissivity correction for interpreting thermal radiation from a terrestrial surface. J Appl Meteorol 18:1165–1171
Tomlinson CJ et al (2011) Remote sensing land surface temperature for meteorology and climatology: a review. Meteorol Appl 18:296–306
van Leeuwen TT et al (2011) Optimal use of land surface temperature data to detect changes in tropical forest cover. J Geophys Res 116:G02002. doi:10.1029/2010JG001488

Weng Q, Lu D, Schubring J (2004) Estimation of land surface temperature-vegetation abundance relationship for urban heat island studies. Remote Sens Environ 89:467–483

Weng Q, Yang S (2006) Urban air pollution patterns, land use and thermal landscape: an examination of the linkage using GIS. Environ Monit Assess 117:463–489

Wu C, Niu Z (2012) Modelling light use efficiency using vegetation index and land surface temperature from MODIS in Harvard forest. Int J Remote Sens 33:2261–2276

Wukelic GE et al (1989) Radiometric calibration of landsat thermatic mapper thermal band. Remote Sens Environ 28:339–347

Yue W et al (2007) The relationship between land surface temperature and NDVI with remote sensing: application to Shanghai Landsat 7 ETM+ data. Int J Remote Sens 28:3205–3226

Geospatial Strategy for Estimation of Soil Organic Carbon in Tropical Wildlife Reserve

Gargi Gupta, Jyoti Singh, Prem Chandra Pandey, Vandana Tomar, Meenu Rani and Pavan Kumar

Abstract This study focuses on the estimation of soil organic carbon of Sariska Wildlife Reserve. The soil organic carbon is one of the most important issues in the research area of the global carbon cycle as it is the largest terrestrial carbon pool. Geospatial and various forest inventory approaches were used during study for statistical correlation between estimated and predicted value. Remote sensing plays a vital role in spatial data acquisition of the ecosystem carbon dynamics at local, regional, and global scale. The advantage of remote sensing is that it provides synoptic observation, periodical and continuous measurement, and availability of digital data for processing standardization. IRS P6 LISS III data (September 2012) were used to analyze the precise estimation of the percentage of the soil organic carbon associated with organic matter in soil. Statistical analysis was performed for finding the regression curve between the predicted and estimated value of soil

G. Gupta · J. Singh · V. Tomar · P. Kumar (✉)
Department of Remote Sensing, Banasthali University, Newai, Rajasthan, India
e-mail: pawan2607@gmail.com

G. Gupta
e-mail: gargi2590@gmx.com

J. Singh
e-mail: jyoti.ud.4u@gmx.com

V. Tomar
e-mail: vandana7232@gmx.com

P. C. Pandey
Department of Geography, University of Leicester, Leicester, LE1 7RH, UK
e-mail: pcp6@le.ac.uk; prem26bit@gmail.com

M. Rani
Project Directorate for Farming System Research, Meerut, Uttar Pradesh, India
e-mail: meenurani06@gmail.com

P. K. Srivastava et al. (eds.), *Remote Sensing Applications in Environmental Research*, Society of Earth Scientists Series, DOI: 10.1007/978-3-319-05906-8_5,

organic carbon. The results illustrated that the determination of coefficient (r^2) between the predicted and estimated SOC values is found to be 0.708.

Keywords Bare soil index · NDVI · Soil organic carbon · Regression analysis

1 Introduction

Carbon plays an important role in the global carbon cycle (Rani et al. 2011a; Tian et al. 2003; Dong et al. 2003; Bellamy et al. 2005). Carbon in the ecosystem is present in both organic as well as inorganic form. When carbon is present in terrestrial ecosystem associated with the soil then it is called as soil organic carbon (SOC), and SOC turns out to be the largest terrestrial carbon pool playing an important part in the global carbon cycle. Organic carbon plays an important role in the crop productivity, soil type, fertility, physical characteristic of soil etc. Hence, soil organic carbon is the amount of carbon present in the organic matter of the soil while soil inorganic carbon is different from SOC which is held in soil minerals as carbonates. Soil organic matter includes the material of the biological origin present in the soil independent of the origin and state of decomposition of soil (Baldock and Skjemstad 1999; Baldock and Broos 2008). The organic carbon in the soil is present in different types like humus, particulate organic matter, and crop residue due to the varying size of the organic matter (Broos and Baldock 2008; GRDC 2009). Typically organic matter contains 60 % carbon; hence a soil containing 1 % SOC contains around 1.7 % organic matter (Bell and Lawrence 2009). Soil organic carbon plays an important part of the global carbon cycle (Chang 2008). Carbon is continually entering and leaving the soil, therefore soil act as both sink and source. For CO_2, CH_4 and NO_2 it acts as a source while as a sink for greenhouse gasses depending on the land use/land cover (Lal 1999).

The amount or level of carbon is not constant and there is always an exchange of carbon between the various ecosystem and atmosphere (IPCC 2003; Houghton 2005; Hese et al. 2005; Ramankutty et al. 2007; Fung et al. 2007). The land use/land cover have a direct effect on the carbon exchange between atmospheric and terrestrial ecosystem (Kumar et al. 2012; Tomar et al. 2013; Sharma et al. 2012; Rani et al. 2011b; Houghton and Hackler 1999; IPCC 2003; Righelato and Spracklen 2007; Meyfroidt and Lambin 2008; Cochrane et al. 1999; Houghton et al. 2000; Hirsch et al. 2004; Davidson et al. 2008; Vargas et al. 2008). Amazon forest has been focused by many studies (Cochrane et al. 1999; Houghton et al. 2000; Hirsch et al. 2004; Davidson et al. 2008; Vargas et al. 2008). Due to global warming it has been predicted that there is loss of carbon from soil (Schimel et al. 1994; McGuire et al. 1995). The global scale of organic carbon that is estimated is between 700 and 3,000 PgC (Bouwman 1990) which is based on the approaches like soil groups (Bohn 1976; Buringh 1984) vegetation groups (Ajtay et al. 1979; Bolin et al. 1979). In India the forest soil organic carbon ranges between 5.3 and

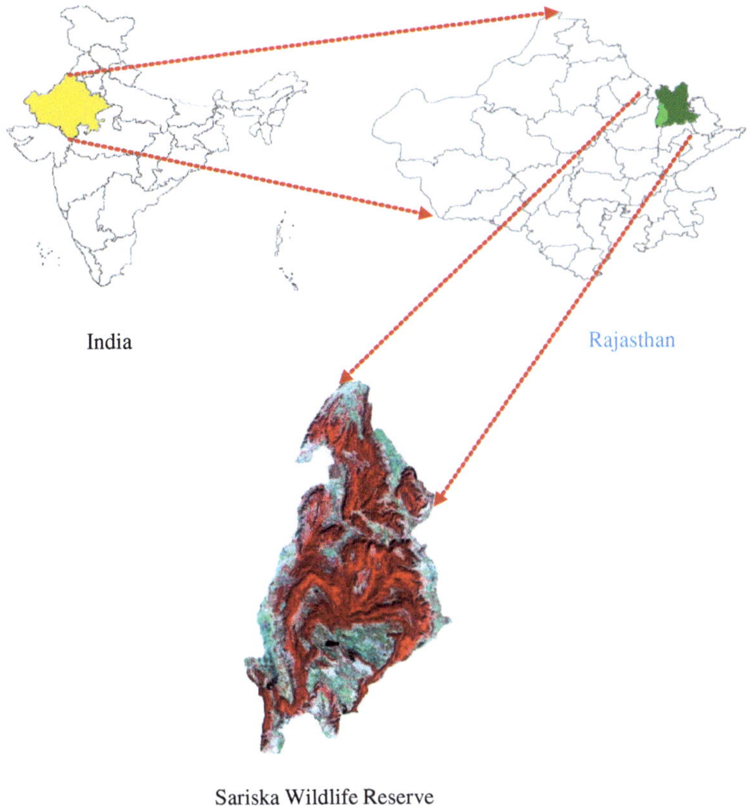

Fig. 1 Location map of the Sariska Wildlife Reserve

6.7 PgC estimated on the basis of global or regional carbon densities (Dadhwal et al. 1998; Ravindranath et al. 1997). The main objective of the study is to analyze the precise estimation of the percentage of the soil organic carbon associated with organic matter in soil and to find the regression curve between the predicted and estimated value of soil organic carbon using statistical analysis.

2 Materials and Methods

2.1 Study Area

The study area (Sariska Wildlife Reserve) lies in the Alwar district, Rajasthan and is located in the older hills of the Aravallies stretched between Mount Abu and Delhi ridge. The study area has geographic extent of 27° 5–27° 33 N latitude and 76° 17–76° 34 E longitudes (Kumar et al. 2013a, b) as shown in Fig. 1. The study

Table 1 Satellite data used in the present study and their specification

Particulars	
Satellite	IRS 1C
Sensor	LISS III
Scale	1:50,000
Band combination	3,2,1
Temporal resolution	5 days
Spatial resolution	23.5 m
Year	September 2012

site covers an area of 866 km^2 including 492 km^2 of the notified Sariska Wildlife Reserve and 374 km^2 of adjoining area of Alwar, Rajgarh and Sariska forest ranges transferred and included in the Sariska Wildlife Reserve. Total forest area under the Sariska Wildlife Reserve consists of 49,199.54 ha in which there are 25 forest blocks, out of which 12 are reserved forest and 13 are preserved forest. The type of the soil and soil characteristics varies according to the type of underlying rocks. The soil type map generated in the study area shows that it contains the soil like loamy, coarse loamy, fine loamy and sandy soil. Major part of the area is covered by the rocks like quartzite, conglomerates, grits, limestone, phyllites, granites and schist. Ancient crystalline and metamorphic rocks with gneisses and schist are generally covered by red sandy soil, which is generally poor in nitrogen, phosphorus and humus contents and are alkaline in nature. There are comparatively rich fertile and dark colored soils in plants and valleys. The soils resulting from the weathering of schistose rocks vary from sandy to heavy loam depending upon the amount of quartzite present in the parent rocks.

2.2 Data Used

The satellite and ancillary data used in the study includes auxiliary and primary data. Primary data were obtained during the field visits by surveying and measuring the study area. The IRS P6 Linear Imaging Self Scanning Sensor III satellite images were used in the study for soil carbon. The specifications of LISS III sensor are illustrated in Table 1.

2.3 Image Interpretation

In the present study, satellite imagery and toposheet acquired from the survey of India (SOI) were used. Digital preprocessing techniques were applied for identifying the different classes and feature in the imagery. The rectification and geo-referencing of satellite imagery was performed using ERDAS IMAGINE 2011©software with reference to toposheet. The 4 bands corresponding to Near

Infra-Red (NIR), Red, Green and Blue of LISS III imagery helped in identifying image characteristics and ground features after image processing technique (Kumar et al. 2013a, b; Kumar and Tomar 2013). Ground truthing of the study area helped out in identifying the different features and land use/land cover correctly.

2.4 Estimation of Soil Organic Carbon

Plot layout during field investigation needs a lot of precautions for collecting samples. We kept in mind all major precautions to be taken during plot layout in the field. Firstly, on hills chooses any aspect and all the plots need to be in same aspect; Second, avoid any major stream inside the plot i.e. the plot can be adjusted to that plot is laid on one side of the stream within the frame of 31.62 × 31.62 m; Third, we need to fix the direction/side of plot, i.e. left or right side of the direction, which is followed uniformly; Fourth precaution is to keep all the baggage outside the proposed plot. This is important to avoid trampling of the herbaceous flora of that corner. Organic soil carbons in each soil sample (humus 0–15 cm and mineral 15–30 cm) were determined using the procedure given by Walkley-Black (1934). In this reaction, carbon is oxidized by the dichromate ion. The excess dichromate ion is then back titrated with ferrous ion (Walkley 1947; Jackson 1958). First of all, 30 soil samples were collected from the Sariska Wildlife Reserve. The sampler chosen in this study is having 3.8 cm radius with soil depth of 5 cm. Every soil sample taken varies in their weight (gm) with a heavy bulk density of approx. 1.4 g/m^3.

During laboratory analysis, 1.00 g soil was taken into 500 ml Erlenmeyer flask. Thereafter, 10 ml of 1N potassium dichromate solution was added to it. Further, 20 ml of sulfuric acid was added to it and mixed by gentle rotation for 1 min, taking care to avoid throwing soil up onto the sides of the flask. Then, it was allowed to stand for 30 min. The solution was diluted to 200 ml with deionized water. Again, 10 ml phosphoric acid, 0.2 g ammonium fluoride, and 10 drops diphenylamine indicator were added to the solution for additional test. The solution was titrated with 0.5N ferrous ammonium sulfate solution until the color changes from dull green to a turbid blue. The titrating solution was added drop by drop until the end point is reached when the color shifts to a brilliant green. Similarly, blank control sample was prepared in the same manner. One duplicate sample and one quality control sample was prepared with each set of samples analyzed. The percentage soil organic carbon is shown in Table 2.

$$\%\ Soil\ Organic\ Carbon = 10[1(S \div B)] \times 0.67 \quad (1)$$

where S = Sample titration, and B = Blank titration

Table 2 Soil organic carbon

Plot no.	Sampler radius (cm)	Soil weight (gm)	Bulk density (g/m^3)	% SOC	SOC (t/ha)
1	3.8	360	1.58	0.68	5.396
2	3.8	320	1.41	0.98	6.912
3	3.8	342	1.50	1.1	8.292
4	3.8	348	1.53	1.12	8.591
5	3.8	347	1.52	1.21	9.255
6	3.8	362	1.59	1.23	9.815
7	3.8	322	1.41	1.32	9.369
8	3.8	330	1.45	1.26	9.165
9	3.8	306	1.34	1.21	8.161
10	3.8	315	1.38	1.32	9.165
11	3.8	300	1.32	1.6	10.580
12	3.8	344	1.51	1.58	11.981
13	3.8	325	1.43	1.6	11.462
14	3.8	356	1.56	1.79	14.047
15	3.8	345	1.52	1.86	14.145
16	3.8	349	1.53	1.96	15.078
17	3.8	351	1.54	1.98	15.319
18	3.8	349	1.53	2.01	15.463
19	3.8	342	1.50	2.06	15.530
20	3.8	352	1.55	2.08	16.139
21	3.8	312	1.37	2.11	14.511
22	3.8	326	1.43	2.13	15.306
23	3.8	356	1.56	2.13	16.715
24	3.8	321	1.41	2.2	15.567
25	3.8	312	1.37	2.2	15.130
26	3.8	324	1.42	2.2	15.712
27	3.8	326	1.43	2.3	16.528
28	3.8	365	1.60	2.62	21.080
29	3.8	363	1.60	2.43	19.444
30	3.8	398	1.75	2.54	22.284

3 Results and Discussion

3.1 Land Use Land Cover Classification

Land use is the area which is being used by men for its use and land cover is the area where natural resources have no disturbance or human interference. Land use/ Land cover (LU/LC) was prepared which was validated with the accurate GPS points taken during field visits. The leica GPS was used to acquire the positions of the point samples, having accuracy of 10 m. The GPS point samples were post processed to match the accuracy with imagery. In the present study, the site classified as land cover includes dense forest, degraded forest, open forest, non forest while land use includes water body, settlement, road and drainage (Fig. 2).

Fig. 2 LU/LC map of the study area

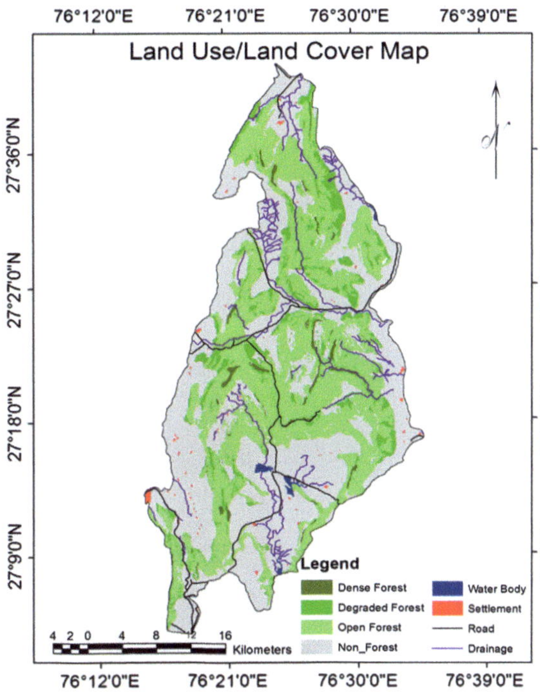

Table 3 Calculated area of Sariska Wildlife Reserve LU/LC classes

LU/LC classes	Area (km^2)	Area (%)
Dense forest	14.191	1.601
Open forest	468.467	52.856
Degraded forest	97.708	11.024
Non forest	292.353	32.986
Settlement	4.293	0.484
Water body	9.293	1.049
Total area	886.305	

The total area under LU/LC was found to be 886.305 km^2. Table 3 shows the area covered by different type's i.e. dense forest, open forest and degraded forest, occupying 1.6, 52.85 and 11.02 % of the area respectively. The lowest area was covered by settlement (0.484 %) and water body (1.049 %).

3.2 *Bare Soil Index*

Bare soil can be defined as the soil and sand on the earth's surface not covered by any grass, wood chips, any live ground covers, artificial turfs and similar covering. Bare soil index (BSI) estimates the value of bare soil in the study area as shown in

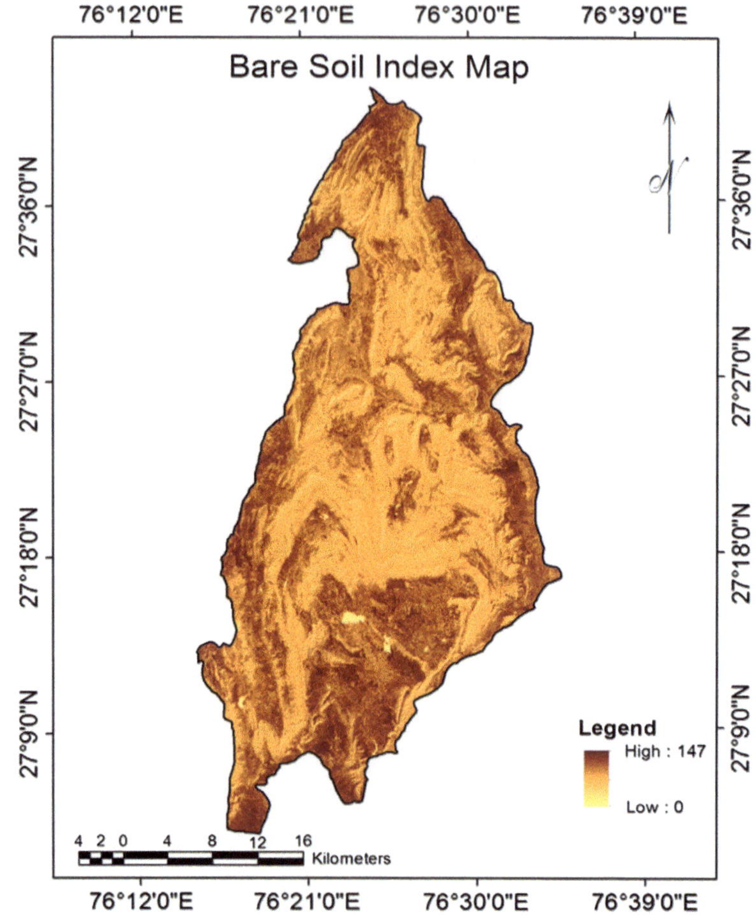

Fig. 3 Bare soil index map of the study area

Fig. 3. BSI helps in separating the vegetation with different background features. Digital classification of satellite data are based on spectral signature and are reported to be more precise. So BSI reduces the effects of bias and assist in the extraction of the significant features of a specific ground object. Hence, the present approach isolates vegetation using BSI indices. The formulae used for calculating the bare soil index is given as (Rikimaru et al. 2002).

$$BSI = [(B5 + B3) - (B4 + B1)]/[(B5 + B3) + (B4 + B1)] \times 100 + 100;\ 0 < BI < 200 \quad (2)$$

The range of BI is converted within 8 bits range.

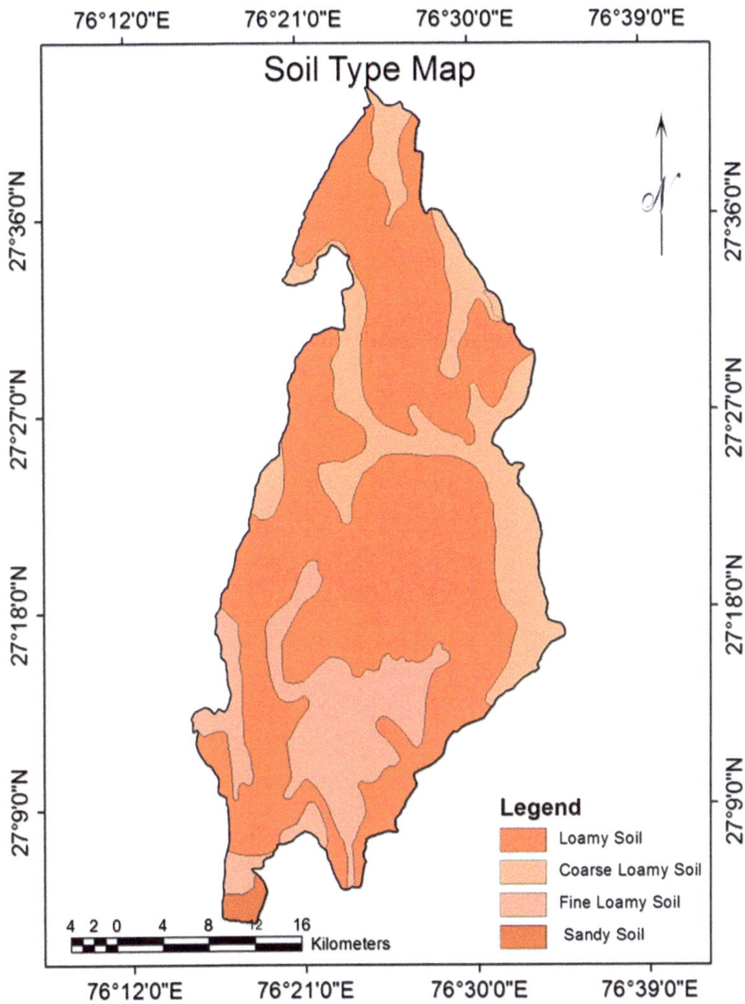

Fig. 4 Soil Type map of the study area

3.3 Soil Type Map

Sariska Wildlife Reserve soil properties are problematic to measure, but can be projected with acceptable accuracy from other soil parameters of the same location. Spectral characteristic pattern of soil is generally overseen by a many number of aspects like soil color, texture, salinity, structure, mineralogy, moisture content, macro and micro organic matter. The literature review designates the major soil properties showing relatively high correlation with remote sensing images. These soil properties are soil moisture, total organic carbon, chemical and physical

Table 4 Estimated and predicted SOC

Plot no.	Predicted SOC	Estimated SOC
1	0.293	5.396
2	0.255	6.912
3	0.233	8.292
4	0.278	8.591
5	0.235	9.255
6	0.279	9.815
7	0.485	9.369
8	0.251	9.165
9	0.351	8.161
10	0.360	9.165
11	0.380	10.580
12	0.448	11.981
13	0.434	11.462
14	0.437	14.047
15	0.444	14.145
16	0.455	15.078
17	0.487	15.319
18	0.456	15.463
19	0.466	15.530
20	0.471	16.139
21	0.458	14.511
22	0.480	15.306
23	0.478	16.715
24	0.490	15.567
25	0.490	15.130
26	0.486	15.712
27	0.480	16.528
28	0.506	21.080
29	0.515	19.444
30	0.530	22.284

properties of soil, salinity clay, silt and sand contents. The soil type map was prepared using ground data and GIS with reference to National Bureau of Soil Survey shown in Fig. 4.

As the above figure depicts about the extent and area covered by the soil type, we can see that the loamy soil covers a large part of the area in the core heart whereas the sandy soil covers least only in the south region. The other two types lies in buffer region such that coarse loamy has its higher percentage in north and north–west region and fine loamy in south parts.

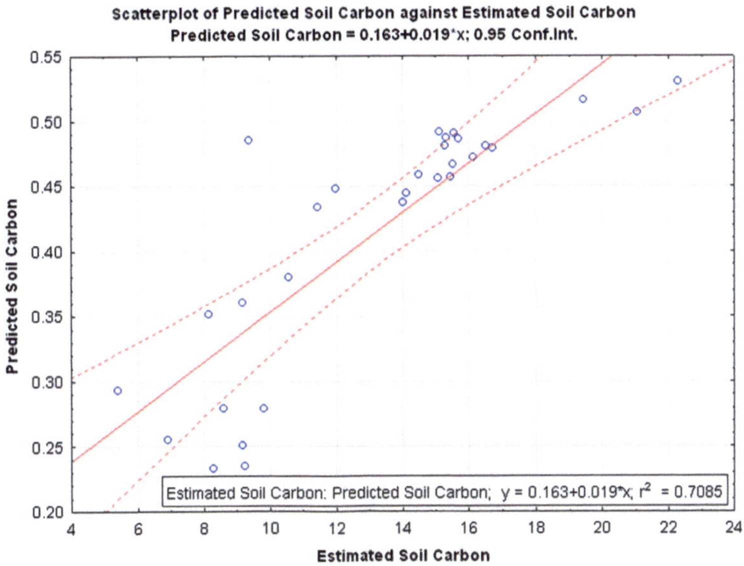

Fig. 5 Regression analysis between estimated and predicted SOC

3.4 Soil Organic Carbon and Regression Analysis

In the present study, R^2 values are used for determination of the linear model. R^2 is the coefficient of determination which has long been used to compare the different models in past study. It was observed that linear equation gives the results which is 0.708. It was observed that there is a positive correlation between the estimated and predicted value. Total carbon is determined using band data and NDVI generated using band data (Table 4). The determination of coefficient (r^2) between the predicted and estimated SOC values is found to be 0.708 (Fig. 5). Thus, carbon maps (Fig. 6) are generated using linear equations.

$$SOC = 0.163 + 0.019 \times NDVI \tag{3}$$

where SOC = Soil organic carbon, and NDVI = Normalized Difference Vegetation Index.

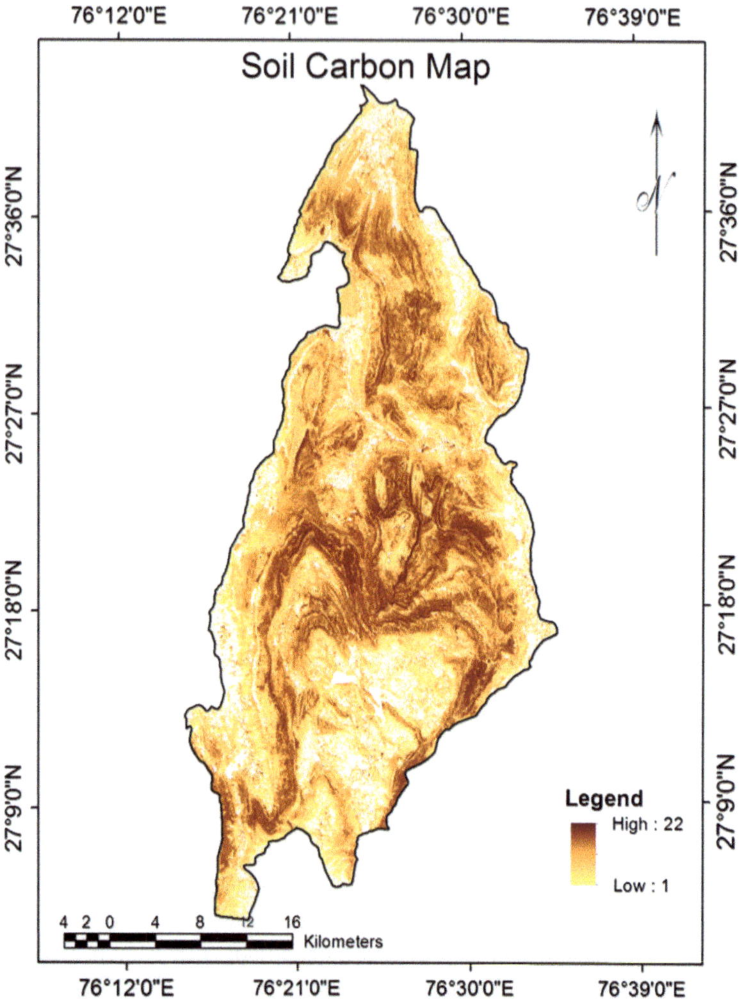

Fig. 6 Soil Carbon map of the study area

4 Conclusion

Linear model was used in the present study to determine the soil condition of the region. R^2 (coefficient of determination) has long been used to compare models for forest parameters like tree volume, above ground biomass, leaf area index and heights (Segura and Kanninen 2005; Lu et al. 2012; Zhao et al. 2009; Samalca 2007). However, for models with different set of variables, R^2 gives misleading results. Total carbon can be determined forming spectral modeling using band data and NDVI generated from the band data. Carbon maps are generated using the linear equations. The estimated values generated from the field study correlates with the predicted value generated from spectra spectral modeling.

This collected data containing the information on location, soil type, texture, measured/estimated bulk density helps in estimating the soil organic carbon present in the soil of Sariska Wild life reserve for depth of 5 cm. The estimated SOC densities were combined with the remote sensing. The study exposes the approaching of LISS III image in estimating SOC for the heterogeneous tropical forest. The extraordinary positive correlation between the estimated SOC directly from field parameters and predicted SOC from spectral band information demonstrates the fact that NDVI can be considered to be an effective spectral vegetation index to estimate SOC. Linear models also showing comparable results and can be considered as standard.

Acknowledgment We are thankful to Haryana Space Applications Centre (Department of Science and Technology), Hisar, Haryana, India for providing satellite and ancillary data support. We are thankful to Department of Remote Sensing, Banasthali Vidyapith, Rajasthan-304022, India for providing all necessary support.

References

Ajtay GL, Ketner P, Duvigneaud P (1979) Terrestrial primary production and phytomass. In: Bolin B, Degens ET, Kempe S, Ketner P (eds) The global carbon cycle (SCOPE 13). Wiley, Chichester, pp 129–182

Baldock JA, Skjemstad JO (1999) Soil organic carbon/soil organic matter. In: Peverill KI, Sparrow LA, Reuter DJ (eds) Soil analysis: an interpretation manual, pp 159–170

Baldock J, Broos K (2008) Can we build-up carbon and can we sell it? Australian Grain, 4–9

Bell M, Lawrence D (2009) Soil carbon sequestration-myths and mysteries, Queensland Department of Primary Industries and Fisheries, p. 7. http://www.futurebeef.com.au/wp-content/uploads/2011/09/Soil-Carbon-Sequestration-Report.pdf. Accessed 20 Aug 2013

Bellamy PH, Loveland PJ, Bradley RI, Lark RM, Kirk GJD (2005) Carbon losses from all soils across England and Wales 1978–2003. Nature 437:245–248

Bohn HL (1976) Estimate of organic carbon in world soils II. Soil Sci Soc Am J 40(3):468–470

Bolin B, Degens ET, Duvigneaud P, Kempe S (1979) The global biogeochemical carbon cycle. In: Bolin B, Degens ET, Kempe S, Ketner P (eds) SCOPE 13. Wiley, Chichester, pp 1–56

Bouwman AF (1990) Soils and the greenhouse effect. In: Proceedings of the international conference on soils and the greenhouse effect. Wiley, New York

Broos K, Baldock J (2008) Building soil carbon for productivity and implications for carbon accounting, in 2008 South Australian GRDC Grains Research Update

Buringh P (1984) Organic carbon in the soils of the world. In: Woodwell GM (ed) The role of terrestrial vegetation in the global carbon cycle. Measurement by remote sensing, SCOPE 23. Wiley, New York, pp 91–109

Chang Y (2008) Increasing soil organic carbon of agricultural land-Prime Fact 735, NSW Department of Primary Industries

Cochrane MA, Alencar A, Schulze MD, Souza CM Jr, Nepstad DC, Lefebvre P, Davidson EA (1999) Positive feedbacks in the fire dynamic of closed canopy tropical rain forests. Science 284:1832–1835

Dadhwal VK, Pandya N, Vora AB (1998) Carbon cycle for Indian forest ecosystem: a preliminary estimated. In: Subbaraya BH, Rao DP, Desai PS, Manikiam, Rajaratnam P (eds) Global change studies: scientific Results from ISRO-GBP. ISRO, Bangalore, pp 411–430

Davidson EA, de Abreu Sá TD, Carvalho CJR, de Oliveira Figueiredo R, do Socorro A, Kato M (2008) An integrated greenhouse gas assessment of an alternative to slash-and-burn agriculture in eastern Amazonia. Glob Change Biol 14:998–1007

Dong J, Kaufmann RK, Myneni RB, Tucker CJ, Kauppi PE, Liski J, Buermann W, Alexeyev V, Hughes MK (2003) Remote sensing estimates of boreal and temperate forest woody biomass: carbon pools, sources, and sinks. Remote Sens Environ 84:393–410

Fung IY, Doney SC, Lindsay K, John J (2007) Evolution of carbon sinks in a changing climate. Proc Nat Acad Sci 102:11201–11206

GRDC (2009) Carbon farming (fact sheet), Grains Research and Development Corporation

Hese S, Lucht W, Schmullius C, Barnsley M, Dubayah R, Knorr D, Neumann K, Riedel T, Schröter K (2005) Global biomass mapping for an improved understanding of the CO_2 balance-the Earth observation mission Carbon-3D. Remote Sens Environ 94:94–104

Hirsch AI, Little WS, Houghton RA, Scott NA, White JD (2004) The net carbon flux due to deforestation and forest re-growth in the Brazilian Amazon: analysis using a process-based model. Glob Change Biol 10:908–924

Houghton RA (2005) Aboveground forest biomass and the global carbon balance. Glob Change Biol 11:945–958

Houghton RA, Hackler JL (1999) Emissions of carbon from forestry and land-use change in the tropical Asia. Glob Change Biol 5:481–492

Houghton RA, Scole DL, Nobre CA, Hackler JL, Lawrence KT, Chomentowskl WH (2000) Annual fluxes of carbon from deforestation and regrowth in the Brazilian Amazon. Nature 403:301–304

IPCC National greenhouse gas inventories programme (2003) Good practice guidance for land use. Land-use change and forestry institute for global environmental strategies. Hayama, pp 1–275

Jackson ML (1958) Soil chemical analysis. Prentice Hall, New Jersey, pp 214–221

Kumar P, Kumar D, Mandal VP, Pandey PC, Rani M, Tomar V (2012) Settlement risk zone recognition using high resolution satellite data in Jharia coal field, Dhanbad. India Life Sci J 9(1s):1–6

Kumar P, Singh B, Rani M (2013a) An efficient hybrid classification approach for land use/land cover analysis in semi-desert area using ETM+ and LISS-III sensor. Sens J IEEE 13(6):2161–2165

Kumar P, Sharma LK, Pandey PC, Sinha S, Nathawat MS (2013b) Geospatial strategy for forest biomass estimation of tropical forest of Sariska Wildlife Reserve (India). IEEE J Sel Top Appl Earth Obs Remote Sens 6(2):917–923

Kumar P, Tomar V (2013) Monitoring of traffic and its impact on environment using geospatial technology. J Ecosyst Ecogr 3:123. doi:10.4172/2157-7625.1000123

Lal R (1999) Soil management and restoration for C sequestration to mitigate the accelerated greenhouse effect. Prog Environ Sci 1(4):307–326

Lu D, Chen Q, Wang G, Moran E, Batistella M, Zhang M, Laurin GV, Saah D (2012) Above ground forest biomass estimation with landsat and LiDAR data and uncertainty analysis of the estimates. Int J Forest Res 2012:16. doi:10.1155/2012/436537

McGuire AD, Mellilo JM, Kicklighter DW, Joyce LA (1995) Equilibrium responses of soil carbon to climate change: empirical and processes based estimates. J Biogeogr 22:785–796

Meyfroidt P, Lambin EF (2008) Forest transition in vietnam and its environmental impacts. Glob Change Biol 14:1–18

Ramankutty N, Gibbs HK, Achard F, Defries R, Foley JA, Houghton RA (2007) Challenges to estimating carbon emissions from tropical deforestation. Glob Change Biol 13:51–66

Rani M, Kumar P, Yadav M, Hooda RS (2011a) Wetland assessment and monitoring using image processing technique: a case study of Ranchi, India. J Geogr inf syst 3:345–350

Rani M, Kumar P, Yadav M, Hooda RS (2011b) Role of geospatial techniques in forest resource management of Sariska Tiger Reserve (Rajasthan), India. New York Sci J 4(6):77–82
Ravindranath NH, Somashekhar BS, Gadgil M (1997) Carbon flows in Indian forests. Clim Change 35:297–320
Righelato R, Spracklen DV (2007) Carbon mitigation by biofuels or by saving and restoring forests. Science 317:902
Rikimaru A, Roy PS, Miyatake S (2002) Tropical forest cover density mapping. Trop Ecol 43(1):39–47
Samalca IK (2007) Estimation of forest biomass and its error. A case in Kalimantan, Indonesia. Master's thesis. International Institute of Geo-Information Science and Earth Observation, Enschede, the Netherlands
Schimel DS, Braswell BH, Holland EA, McKeowan R, Ojima DS, Painter Th, Patron WJ, Townshend AR (1994) Climate, edaphic and biotic controls over storage and turnover of carbon in soils. Global Biogeochem Cycles 8:279–293
Segura M, Kanninen M (2005) Allometric models for the tree volume and total aboveground biomass in a tropical humid forest in Costa Rica. Biotropica 37(1):2–8
Sharma LK, Pandey PC, Nathawat MS (2012) Assessment of land consumption rate with urban dynamic changes using geospatial approach. J Land Use Sci 7(2):131–148
Tomar V, Kumar P, Rani M, Gupta G, Singh J (2013) A satellite-based biodiversity dynamics capability in tropical forest. Electron J Geotech Eng 18:1171–1180
Tian H, Melillo JM, Kicklighter DW, Pan S, Liu J, McGuire AD, Moore B III (2003) Regional carbon dynamics in monsoon Asia and its implications for the global carbon cycle. Global Planet Change 37:201–217
Vargas R, Allen MF, Allen EB (2008) Biomass and carbon accumulation in a fire chronosequence of a seasonally dry tropical forest. Glob Change Biol 14:109–124
Walkley A (1947) A critical examination of a rapid method for determination of organic carbon in soils-effect of variations in digestion conditions and of inorganic soil constituents. Soil Sci 63:251–257
Walkley A, Black IA (1934) An examination of Degtjareff Method for determining soil organic matter and a proposed modification of the chromic acid titration method. Soil Sci 37:29–37
Zhao K, Popescu S, Nelson R (2009) Lidar remote sensing of forest biomass: a scale-invariant estimation approach using airborne lasers. Remote Sens Environ 113:182–196

Part II
Advanced Remote Sensing Applications

A Comparative Assessment Between the Application of Fuzzy Unordered Rules Induction Algorithm and J48 Decision Tree Models in Spatial Prediction of Shallow Landslides at Lang Son City, Vietnam

Dieu Tien Bui, Biswajeet Pradhan, Inge Revhaug and Chuyen Trung Tran

Abstract The main objective of this study is to investigate potential application of the Fuzzy Unordered Rules Induction Algorithm (FURIA) and the Bagging (an ensemble technique) in comparison with Decision Tree model for spatial prediction of shallow landslides in the Lang Son city area (Vietnam). First, a landslide inventory map was constructed from various sources. Then, the landslide inventory was randomly partitioned into 70 % for training the models and 30 % for the model validation. Second, six landslide conditioning factors (slope, aspect, lithology, land use, soil type, and distance to faults) were prepared. Using these factors and the training dataset, landslide susceptibility indexes were calculated using the FURIA, the FURIA with Bagging, the Decision Tree, and the Decision Tree with Bagging. Finally, prediction performances of these susceptibility maps were carried out using the Receiver Operating Characteristic (ROC) technique. The results show that area under the ROC curve (AUC) using training dataset has the largest for the Decision Tree with Bagging (0.925) and the FURIA with Bagging (0.913), followed by the Decision Tree (0.908) and the FURIA (0.878). The prediction capability of these

D. Tien Bui (✉) · I. Revhaug
Department of Mathematical Sciences and Technology, Norwegian University of Life Sciences, P.O. Box 5003IMT, N-1432 Aas, Norway
e-mail: buitiendieu@gmail.com

D. Tien Bui
Faculty of Surveying and Mapping, Hanoi University of Mining and Geology, Dong Ngac, Tu Liem, Hanoi, Vietnam

C. Trung Tran
Faculty of Information Technology, Hanoi University of Mining and Geology, Dong Ngac, Tu Liem, Hanoi, Vietnam

B. Pradhan
Faculty of Engineering, Department of Civil Engineering, University Putra Malaysia, 43400 Serdang, Selangor Darul Ehsan, Malaysia

P. K. Srivastava et al. (eds.), *Remote Sensing Applications in Environmental Research*, Society of Earth Scientists Series, DOI: 10.1007/978-3-319-05906-8_6,

models was estimated using the validation dataset. The highest prediction was achieved using the FURIA with Bagging (AUC = 0.802), followed by the Decision Tree (AUC = 0.783), the Decision Tree with Bagging (AUC = 0.777), and the FURIA (AUC = 0.773). We conclude that the FURIA with Bagging is the best model in this study.

Keywords GIS · Landslide susceptibility · Remote sensing · FURIA · Decision tree

1 Introduction

Landslides are one of many types of natural processes and when threaten mankind they will represent as hazard (Glade et al. 2005). Globally, landslides cause thousands of deaths and injuries, and the direct and indirect costs of landslide damages go up to many billions of USD annually (Roberds 2005). Climate changes and its anticipated consequences are expected to lead to an increase in natural hazards including landslides, resulting in loss of lives and infrastructure damages (Korup et al. 2012).

Landslide damages can be reduced if we understand the mechanisms of occurrence, prediction, hazard assessment, early warning, and risk management (Sassa and Canuti 2008). Landslide hazard assessment can help authorities to reduce landslide damages through proper land use management for infrastructural development and for environmental protection (Tien Bui et al. 2013a). The spatial prediction of landslides is considered as one of the most difficult aspects in the assessment of landslide hazard. For this reason, various methods and techniques have been proposed and they range from simple qualitative techniques to sophisticated mathematical models (Chung and Fabbri 2008). Good overview of these methods including their disadvantages and advantages can be seen in Guzzetti et al. (1999) and Chacon et al. (2006).

In recent years, with the development of geographical information systems (GIS) and computer sciences, some new methods such as neural networks, fuzzy logic, and neuro-fuzzy have become new solutions for landslide modelling with good prediction capabilities (Pradhan et al. 2010; Sezer et al. 2011; Pourghasemi et al. 2012; Tien Bui et al. 2012c; Akgun et al. 2012; Althuwaynee et al. 2014). Although a series of methods and techniques have been proposed and implemented, no agreement has been reached so far on which method is the best one for landslide susceptibility mapping. It is clear that the quality of landslide susceptibility models is influenced both by the methods used and the sampling strategies employed. In more recent years, data mining and ensemble-based approaches have received much attention in many fields including landslide studies (Tien Bui et al. 2013b). They are reported having an improvement of the prediction performance of models (Rokach 2010; Tien Bui et al. 2013c).

The main objective of this study is to investigate potential application of the Fuzzy Unordered Rules Induction Algorithm (FURIA) with Bagging (an ensemble technique) in comparison with the Decision Tree model, for spatial prediction of shallow landslides at Lang Son city area (Vietnam). FURIA is a fuzzy rule based classification system that combines advantages of RIPPER (Cohen 1995) and fuzzy logic. FURIA and its ensemble have not been used in landslide modelling. The computation process was carried out using WEKA ver.3.6.6, MATLAB 7.11, and ArcGIS 10.

2 Study Area and Spatial Database

2.1 Study Area Characteristics

The study area that includes the Lang Son city and the Dong Dang town (Fig. 1) is located in the northeast mountainous province of Lang Son (Vietnam). It covers an area of about 168 km^2 and lies between longitudes 106°41′34″E and 106°48′32″E, and latitudes 21°49′43″N and 21°57′13″N. Slopes in the study area are from 0° to 84°, around 66 % of the study area has slopes steeper than 15°. The elevation ranges from 194 to 800 m a.s.l with a mean of 328 m.

The study area is comprised of approximately 45.2 % forest land, 21.5 % paddy land, 20.4 % barren land, and 5.7 % crop land, whereas settlement areas cover about 6.9 %. The soil types are mainly ferralic acrisols, dystric gleysols, rhodic ferralsols, and eutric fluvisols that account for 95.2 % of the total study area. Eleven lithologic formations are recognized in the region and six of them account for 80 % of the study area. They are Na Khuat, Tam Lung, Khon Lang, Lang Son, Tam Danh, and Mau Son formations. The main lithologies are marl, siltstone, tuffaceous conglomerate, griltstone, sandstone, basalt, and clay shale. Approximately 16 % of the study area is covered by Quaternary deposits that mainly contain granule, grit, breccia, boulder, sand, and clay.

Landslides in the study area mainly occurred during extreme rainfall events and tropical rainstorms. With the rapid development of economics in the province for the last two decades, the expansions of the infrastructures and the settlements which are shifted into the mountainous regions, have increased slope disturbance. In addition, the deforestation is still continuing leading to potential increase of landslides.

2.2 Spatial Database

2.2.1 Landslide Inventory

In the study area, the landslides were mainly rainfall-triggered shallow soil slides and debris flows. Rock fall was reported in some very few cases and is not included in this study. No information on earthquake-induced landslides has been reported so far. The landslide inventory map for this study was constructed from several sources:

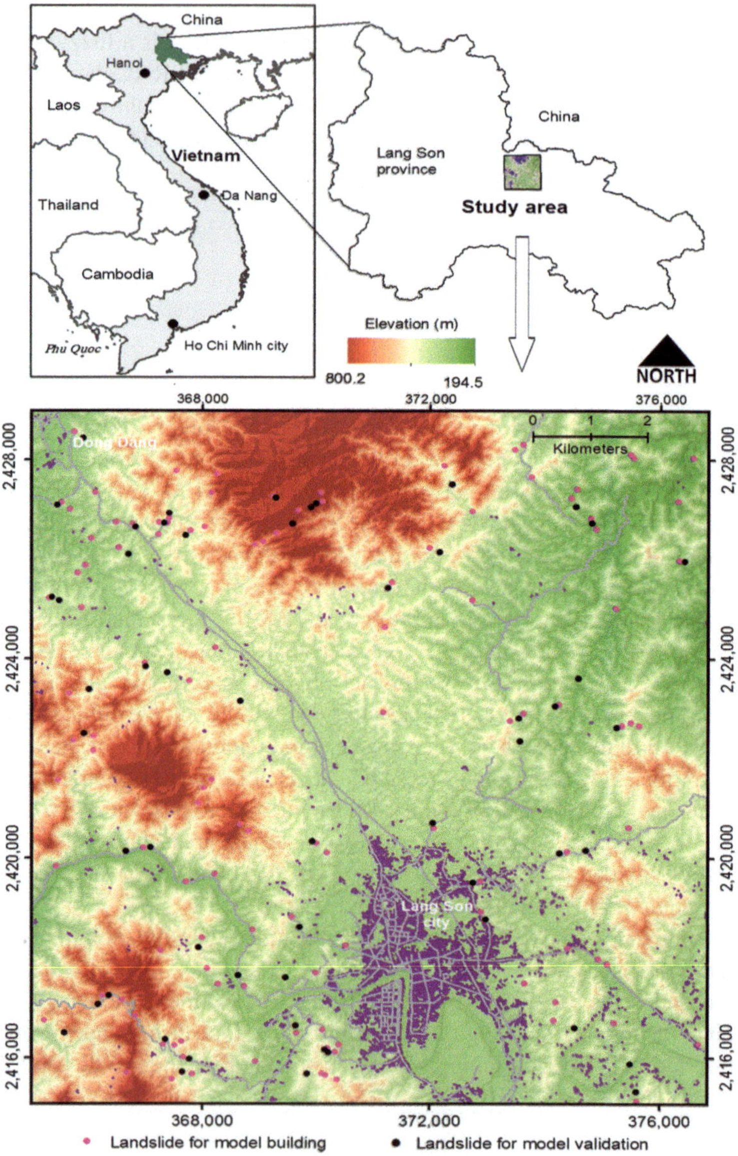

Fig. 1 Study area location map showing the landslide inventory

(1) Landslides that occurred before the year of 2003 detected by the interpretation of aerial photographs and field survey data. The aerial photographs have a resolution of about 1 m. The aerial photographs were acquired by the Aerial Photo—Topography Company 2003; (2) the landslide inventory map of 2006 (Tam et al. 2006); (3) the landslide inventory map of 2009 (Truong et al. 2009); (4) Some recent landslides

were identified during field works. A total of 172 landslides depicted by polygons (Fig. 1) were identified and registered in the inventory map, including 86 rotational slides, 52 translational slides, and 34 debris flows.

2.2.2 Digital Elevation Model and Derivatives

In this study area, the digital elevation model (DEM) was generated from National Topographic Maps at scales 1:5,000 for the Lang Son city and 1:10,000 for the surrounding areas. The DEM has 5 m resolution. Slope and aspect were extracted from the DEM. In the case of the slope map, six categories were constructed (Fig. 2a), whereas nine layer classes were constructed for the aspect map (Fig. 2b).

2.3 Lithology and Distance to Faults

The lithological map was constructed with seven groups: conglomerate, basalt, quaternary, siltstone, limestone, sandstone, and tuff (Fig. 2c). The distance-to-faults map (Fig. 3b) was constructed by buffering the fault lines. Five fault buffer categories were constructed: 0–100, 100–200, 200–300, 300–400, and >400 m. The lithology and fault lines were extracted from four tiles of the Geological and Mineral Resources Map of Vietnam at 1:50,000 scale (Quoc et al. 1992; Truong et al. 2009).

2.4 Land Use and Soil Type

Land use was extracted from the land use status map from 2010 of the Lang Son province. The scale of the land use status map is 1:50,000 and this map is a result of the Status Land Use Project of the National Land Use Survey in Vietnam. A total of nine classes were constructed for the land use map (Fig. 2d). Regarding the soil type map, a total of eight layers were constructed for analysis (Fig. 3a). The soil types were extracted from the national pedology map at scale 1:100,000.

3 Methodology

3.1 Training and Validation Dataset

The landslide inventory and six conditioning factor maps (slope, aspect, lithology, landuse, soil type, and distance to faults) were converted to a grid cell format with spatial resolution of 5 m. Assuming *N*(*LS*) is the total number of grid cells in the

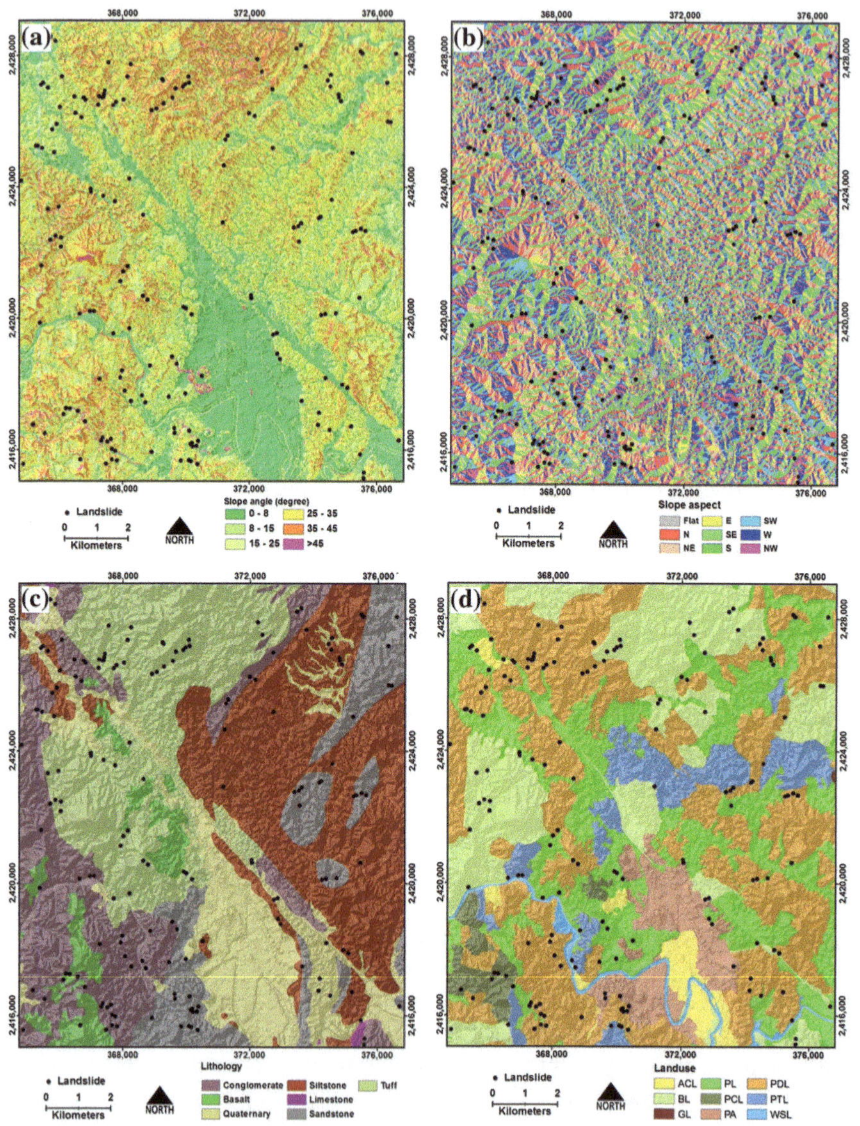

Fig. 2 **a** Slope; **b** Aspect; **c** Lithology; **d** Land use (*ACL* annual crop land; *BL* barren land; *GL* grass land; *Pl* paddy land; *PCL* perennial crop land; *PA* populated area; *PDL* productive forest land; *PTL* protective forest land; *WSL* water surface land)

study area and the training dataset D has $N(D)$ total number of landslide grid cells. We define F_{ij} as the j-th layer class of the landslide conditioning factors F_i and $N(F_{ij})$ is the total number of grid cells in the class F_{ij}. By overlaying the landslide grid cells in the training dataset on each of the six landslide conditioning maps, the

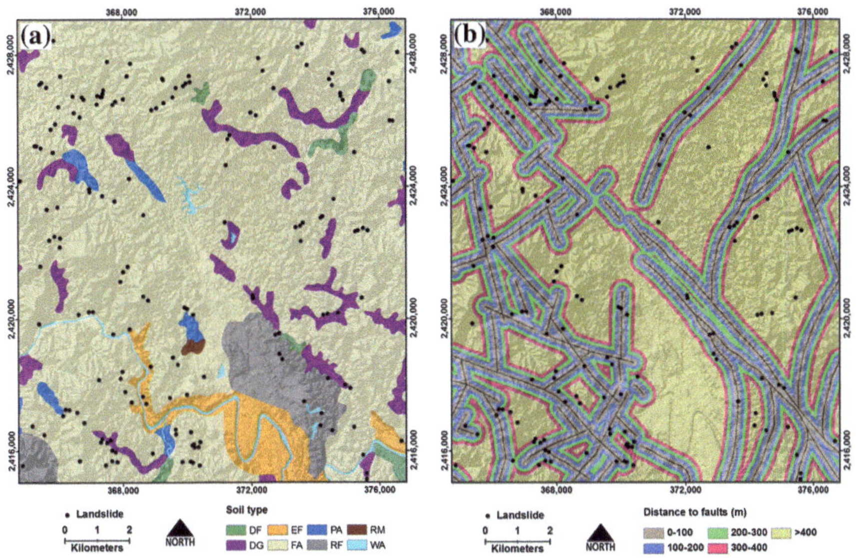

Fig. 3 **a** Soil type; **b** Distance to faults; (*DF* dystric fluvisols; *DG* dystric gleysols; *EF* eutric fluvisols; *FA* ferralic acrisols; *PA* plinthic acrisols; *RF* rhodic ferralsols; *RM* rocky mountain; *WA* water area)

number of grid cells in F_{ij} overlapping with the landslide grid cells $\mathrm{N}(\mathrm{T} \cap \mathrm{F_{ij}})$ was determined. Then, each category of the six maps was assigned to an attribute value that was calculated using the following equation

$$Attribute(F_{ij}) = \frac{W_{F_{ij}}}{\sum_{j=1}^{n} W_{F_{ij}}} \quad (1)$$

where

$$W_{F_{ij}} = \frac{N(D \cap F_{ij})/N(D)}{[N(F_{ij}) - N(D \cap F_{ij})]/[N(LS) - N(D)]} \quad (2)$$

The numerator in Eq. (2) is the proportion of landslide pixels that occur in the factor class, whereas the denominator is the proportion of non-landslide pixels in the factor class.

In landslide susceptibility modeling, a landslide inventory is suggested to be partitioned into two subsets (Chung and Fabbri 2003), one subset will be used for building the landslide models whereas the other will be used for model validation. In general, the partition of landslide inventories using temporal distribution is considered to be the best method (Chung and Fabbri 2008). However, the dates for the past landslide are unknown; therefore we randomly split the landslide inventory map in a 70/30 ratio for building and validation of the model, respectively.

Resulting in a training dataset that contains 117 landslide locations (3,793 landslide grid cells), that was used for building models, and a validation dataset with 55 landslide locations (1,664 landslide grid cells). Landslide pixels were assigned a value of 1.

The same number of grid cells was randomly sampled from the no landslide areas and were assigned a value of 0. A total of 3,793 no-landslide grid cells were generated for the training data and 1,664 no-landslide grid cells for the validation data. At the final step, the values of the six landslide conditioning factors were extracted to build the training and validation datasets. The training and validation datasets contain 7,586 and 3,328 observations, one dependent variable, and six independent variables (the six landslide conditioning factors) (Table 1).

3.2 Fuzzy Unordered Rules Induction Algorithm

FURIA is a fuzzy rule based classification system proposed by Huhn and Hullermeier (2009). This algorithm is an extension of a state-of-the-art rule learner called RIPPER (Cohen 1995) in which fuzzy and unordered rules are to be used instead of conventional rules and rule lists, respectively.

Suppose that we have a training dataset D that have instance-label pairs (x_i, y_i) where i is the i-th training instance, $\boldsymbol{x}_i \in R^n$, and $y_i \in \{1, 0\}$. In the current context, x_i is the vector of input of the six landslide conditioning factors: slope, aspect, lithology, land use, soil type, and distance to faults. The two classes of $\{1, 0\}$ denote landslide and no-landslide pixels. RIPPER divides the training dataset into two subsets a growing set and a pruning set. The first one will be used for growing the rules whereas the second one is used for pruning. At the first step, rule sets will be generated and learned using the growing set. Each rule to be grown by greedily adding antecedents until the rule is satisfied. All possible combinations of landslide conditioning factors were tested and the final one with the highest value of FOIL's Information Gain (IG) (Quinlan and Cameron-Jones 1993) was selected.

$$IG_r = p_r[\log_2(p_r/(p_r + n_r)) - \log_2(p/(p + n))] \tag{3}$$

where p_r and n_r are the number of positive and negative instances cover by the rule, whereas p and n are the number of positive and negative instances cover by the default rule.

For avoiding over-fitting, the rule pruning process was carried out by simplifying the rules. All of the learned antecedents will be pruned if the antecedents maximizing V_r. Finally, the rule optimization process was carried out.

$$V_r = p_r/(p_r + n_r) \tag{4}$$

FURIA combines advantages of RIPPER and fuzzy logic, and the rule order in the rule list is not important and there is no default rule (Trawinski et al. 2011). Rules

Table 1 Attribute classes of landslide conditioning factors used in the FURIA, FURIA with bagging, decision tree, decision tree with bagging

Data layers	Class	Number of pixels in class	Landslide pixels	$W_{F_{ij}}$	$\sum W_{F_{ij}}$	Attribute
Slope (degree)	0–8	39,813,500	0	0.0000	46.0394	0.000
	8–15	17,114,025	39	0.5599	46.0394	0.012
	15–25	35,543,725	796	5.5028	46.0394	0.120
	25–35	47,169,725	1,549	8.0691	46.0394	0.175
	35–45	25,972,525	1,204	11.3908	46.0394	0.247
	>45	2,455,275	205	20.5169	46.0394	0.446
Aspect	Flat (−1)	7,881,175	0	0.0000	45.3072	0.000
	North	18,304,150	33	0.4430	45.3072	0.010
	Northeast	21,121,050	159	1.8497	45.3072	0.041
	East	20,045,575	262	3.2115	45.3072	0.071
	Southeast	19,713,650	959	11.9535	45.3072	0.264
	South	20,360,850	1,326	16.0028	45.3072	0.353
	Southwest	22,3707,00	847	9.3034	45.3072	0.205
	West	20,089,900	198	2.4217	45.3072	0.053
	Northwest	18,181,725	9	0.1216	45.3072	0.003
Lithology	Conglomerate	30,238,950	1,086	8.8247	30.9657	0.285
	Basalt	7,775,900	54	1.7063	30.9657	0.055
	Quaternary	26,912,475	204	1.8625	30.9657	0.060
	Siltstone	45,368,200	1,081	5.8547	30.9657	0.189
	Limestone	248,450	0	0.0000	30.9657	0.000
	Sandstone	18,722,500	597	7.8351	30.9657	0.253
	Tuff	38,802,300	771	4.8823	30.9657	0.158
Land use	Annual crop land	4,499,225	64	3.4952	34.3324	0.102
	Populated area	10,566,150	155	3.6045	34.3324	0.105
	Protective forest land	12,473,925	322	6.3429	34.3324	0.185

(continued)

Table 1 (continued)

Data layers	Class	Number of pixels in class	Landslide pixels	$W_{F_{ij}}$	$\sum W_{F_{ij}}$	Attribute
	Productive forest land	61,029,725	1,325	5.3346	34.3324	0.155
	Paddy land	39,295,850	584	3.6517	34.3324	0.106
	Barren land	33,753,675	1,304	9.4928	34.3324	0.276
	Perennial crop land	3,974,925	39	2.4108	34.3324	0.070
	Water surface land	2,498,475	0	0.0000	34.3324	0.000
	Grass land	36,825	0	0.0000	34.3324	0.000
Soil type	Ferralic acrisols	133,741,975	3,285	6.0353	20.2541	0.298
	Dystric gleysols	10,298,575	90	2.1473	20.2541	0.106
	Plinthic acrisols	2,195,825	4	0.4476	20.2541	0.022
	Water area	1,788,350	0	0.0000	20.2541	0.000
	Dystric fluvisols	2,082,500	0	0.0000	20.2541	0.000
	Eutric fluvisols	8,029,575	220	6.7323	20.2541	0.332
	Rhodic ferralsols	9,744,950	194	4.8916	20.2541	0.242
	Rocky mountain	247,025	0	0.0000	20.2541	0.000
Distance to faults (m)	0–100	28,343,050	837	7.2563	32.2505	0.225
	100–200	25,966,100	1,292	12.2264	32.2505	0.379
	200–300	22,463,775	402	4.3972	32.2505	0.136
	300–400	17,874,600	399	5.4849	32.2505	0.170
	>400	73,481,250	863	2.8858	32.2505	0.089

for each label class were induced separately using one-versus-the rest strategy. FURIA transforms the crisp rules of RIPPER into fuzzy rules using the trapezoidal membership function (Huhn and Hullermeier 2009). In this function, each fuzzy interval is specified by four parameters and is written as $I = (T_1, T_2, T_3, T_4)$.

$$I(x) = \begin{cases} 1 & T_2 \leq v \leq T_3 \\ \frac{v - T_1}{T_2 - T_1} & T_1 \leq v \leq T_2 \\ \frac{T_4 - v}{T_4 - T_2} & T_3 \leq v \leq T_4 \\ 0 & \text{else} \end{cases} \tag{5}$$

For an instance $v_i = (x_{i1}, \ldots, x_{i6})$, the fuzzy membership function can be expressed as

$$\mu(v_i) = \prod_{j=1}^{6} I_j(x_j) \tag{6}$$

The fuzzification of a single antecedent of a rule is only relevant to a subset $D_k \in D$, and then D_k is divided into two subsets D_{k+} and D_{k-}. The quality of the fuzzification is checked to choose the best one using the purity rule criteria as mentioned in Eq. (7)

$$pur = \frac{p_i}{p_i + n_i}; \quad p_i = \sum_{v \in D_{k+}} \mu A_i(v); \quad n_i = \sum_{v \in D_{k-}} \mu A_i(v)\,; \; A_i \in I(x) \tag{7}$$

Fuzzy rules were constructed for class y_i and a certainty degree CD_i for the consequence. The final decision for output is based on the largest V value as

$$V = \sum_{i=1}^{m} \mu_{rule(i)}(v) \, * \, CD_i \tag{8}$$

Finally, the rule generalization procedure is carried out to obtain the final fuzzy rule list. A detailed explanation can be seen in Huhn and Hullermeier (2009, 2010).

FURIA was training using stratified 10-fold cross-validation. First, the training dataset was randomly partitioned into 10-folds of equal size. Then, in each run, 9-folds were used for fitting the model whereas the remaining fold was used to assess the performance. The procedure is repeated ten times and results are averaged. In this study, the fuzzy aggregation operator of the product T-Norm (used as fuzzy AND) was selected to combine rule antecedents. This is because FURIA product was reported significant better than FURIA-min (the minimum of T-Norm) (Huhn and Hullermeier 2010). Since the selection of number folds for the training data used for pruning is significant affecting the model accuracy, a test was therefore performed with different folds versus classification accuracy. The result shows that 4-folds used for pruning and the rest for growing the fuzzy rules have the highest classification accuracy. Other parameters were set as default in WEKA. Finally, the

FURIA model with 45 rules was constructed for landslide susceptibility in this study. The overall accuracy was 84.84 %. The details for the accuracy by class and performance by the FURIA model are shown in Tables 4 and 5.

3.3 Decision Tree

Decision tree classifiers are hierarchical models composed of a root, internal nodes, leaf nodes, and branches, and have been considered one of the most popular classification methods in data mining. The goal of decision tree modeling is to generate a tree structure that contains a set of rules using the training dataset. The tree structure has the capability to predict the output for a new similar dataset with good accuracy. Once a decision tree model is constructed; it can process new data by following a path from the root node to the leaves and values for the new data will be obtained. Since the output for pixels in landslide susceptibility modeling present continuous values, the decision trees are called regression trees. The key advantage of decision trees is that they are easy to construct. In addition, the results from decision trees are readily interpretable with clear information of the contribution of the variables on the model results. However, decision trees do not allow for multiple outputs and are susceptible to noisy data (Zhao and Zhang 2008).

Various algorithms for constructing decision trees have been successfully developed such as classification and regression tree (CART) (Breiman et al. 1984), Chi-square Automatic Interaction Detector decision tree (CHAID) (Michael and Gordon 1997), ID3 (Quinlan 1986), C4.5 (Quinlan 1993), and J48 (Witten and Frank 2005). However the C.45 algorithm has been considered as the fastest algorithm for machine learning with good classification accuracy (Lim et al. 2000). In this study, the J48 algorithm, which is a Java re-implementation of the C4.5 algorithm, was used. The detailed description of the C4.5 algorithm can be seen in Quinlan (1993). Only a short description of decision tree is discussed here. There are two steps in the decision tree construction, the first one is the tree building and the second one is the tree pruning. The first step of the tree building process is to find the input landslide conditioning factor with the highest gain ratio using the training data set, and then select as the first internal node called root node. In the next step, the training dataset was split based on the root values, and sub-notes were created. Then, the gain ratio was estimated for each sub-node. The variable with the highest gain ratio is selected, and the recursive partitioning of the training data set is continued until all instances in the training dataset are assigned to leaf nodes or no remaining variables in which the training data can be further split. In some cases, the resulting tree may be obtained with a large number of branches, and thus the tree may over-fit the training dataset with a perfect classification, but the model has a poor classification performance for a new dataset. Therefore, the tree pruning was carried out by removing unessential nodes but with the classification accuracy still remaining (Breiman et al. 1984).

Table 2 Minimum number of instances per leaf

No.	Minimum number of instances per leaf	Classification accuracy (%)	
		Training dataset	Validation dataset
1	1	88.43	69.35
2	2	88.18	68.66
3	4	87.37	68.78
4	6	86.77	68.90
5	8	85.92	71.06
6	9	85.65	70.97
7	10	85.41	70.64
8	12	85.42	71.00
9	13	85.17	72.09
10	14	85.03	73.29
11	15	84.89	72.71
12	16	84.91	72.45
13	18	84.46	72.29
14	20	84.31	71.51

In this study, the first step in constructing decision tree models is to determine the parameters that influencing the classification accuracy of the resulting tree. The type of pruning is based on sub-tree rising. Laplace smoothing was used here to improve probabilistic estimates at leaves (Tien Bui et al. 2012a, 2013b; Tehrany et al. 2013). A test was carried out to find the most suitable parameters for the study area based on the classification accuracy. The results are shown in Tables 2, 3. The results show that the best values for minimum number of instances per leaf and the confident factor are 14 and 0.15 respectively. The selection number of fold of training data used for reduce-error pruning does not affect the accuracy of the model.

Using the training data set and the above mentioned parameters, decision tree model was trained using with stratified 10-fold cross-validation. The 10-fold cross-validation was preferred to be used in order to ensure that the decision trees generalize beyond the training data (Breiman et al. 1984). Finally, the decision tree model was constructed for landslide susceptibility. The size of the tree is 133. The tree has the root node, 65 internal nodes, and 67 leaves. The classification accuracy is 86.82 %. The more detail of accuracy by class and performance of the decision tree model is shown in Tables 4, 5.

3.4 Bagging

Bagging known as bootstrap aggregation, is one of the earliest ensemble algorithms proposed by Breiman (1996). Bagging is a method that uses bootstrap sampling to generate multiple subsets from the training dataset. Each subset is called a bootstrap sample created by sampling the training dataset of the same size with replacement. In the next step, each of the subset will be used to construct a

Table 3 Confidence factor used for pruning

No.	Confidence factor	Classification accuracy (%)	
		Training dataset	Validation dataset
1	0.05	84.47	72.47
2	0.10	84.79	73.17
3	0.15	85.03	73.29
4	0.20	85.10	73.07
5	0.25	85.04	73.07
6	0.30	85.01	72.86
7	0.35	84.99	72.86
8	0.40	85.05	72.98
9	0.45	85.14	72.98
10	0.50	85.15	72.98
11	0.55	85.18	72.92
12	0.60	85.18	72.92

Table 4 Performance of the FURIA, the FURIA with Bagging, the decision tree, and the decision tree with bagging

No.	Parameters	FURIA	FURIA with bagging	Decision tree	Decision tree with bagging
1	Accuracy (%)	84.84	86.38	86.82	87.50
2	Kappa index	0.697	0.728	0.736	0.750
3	MAE	0.150	0.148	0.211	0.209
4	RMSE	0.361	0.331	0.322	0.308

Table 5 Accuracy assessments by classes of the FURIA, the FURIA with bagging, the decision tree, and the decision tree with bagging

Model	True positive rate (%)	False positive rate (%)	F-measure (%)	Class
FURIA	0.920	0.223	0.859	Landslide
	0.777	0.080	0.837	No-landslide
FURIA with bagging	0.945	0.217	0.874	Landslide
	0.783	0.055	0.852	No-landslide
Decision tree	0.921	0.184	0.875	Landslide
	0.816	0.079	0.861	No-landslide
Decision tree with bagging	0.918	0.168	0.880	Landslide
	0.832	0.082	0.869	No-landslide

classifier based model. Then, the final model is determined by aggregating all the based classifiers (Fig. 4).

Using the training data set, the FURIA with Bagging and the Decision tree with Bagging models were trained. The parameters setting for the above two models are

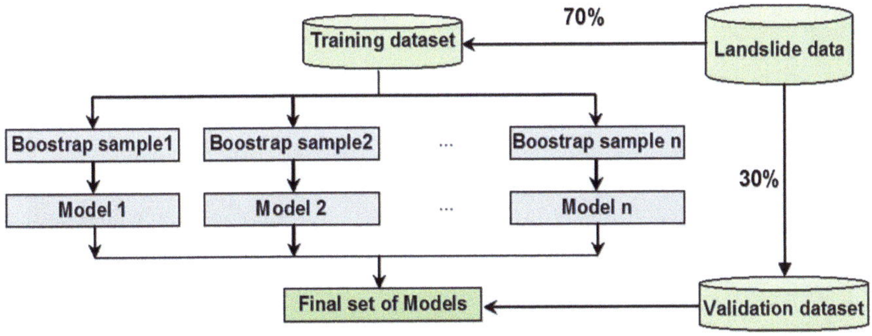

Fig. 4 General framework of the bagged FURIA and the bagged decision tree in this study

remaining the same as in Sects. 3.2 and 3.3. The models were trained and the final results were obtained. The classification accuracy is 86.38 % and 87.50 % for the FURIA with Bagging and for the Decision Tree with Bagging, respectively. The results from the trained models are shown in Tables 4 and 5.

3.5 Generation of Landslide Susceptibility Maps

The successfully trained models were then applied to calculate landslide susceptibility indexes for all the pixels in the study area. The obtained results were converted into a GIS format and loaded in ArcGIS.10. The landslide susceptibility maps were visualized by mean of four susceptibility classes based on the percentage of area (Pradhan and Lee 2010a, b): high (10 %), moderate (10 %), low (20 %), very low (60 %). For the purpose of visualization, only two landslide susceptibility maps that were produced from the FURIA with Bagging and Decision Tree with Bagging models are shown (Figs. 5, 6).

The landslide densities (Kanungo et al. 2008) analysis was carried out for the landslide susceptibility maps by overlaying the four susceptibility zones with the landslide inventory map. Ideally, the density value should increase from very low to high susceptibility zones. The graph of the density analysis for the two models in this study is shown in Fig. 7. The result shows that that there is a gradual increase in landslide density from the very low susceptible zone to the high susceptible zone.

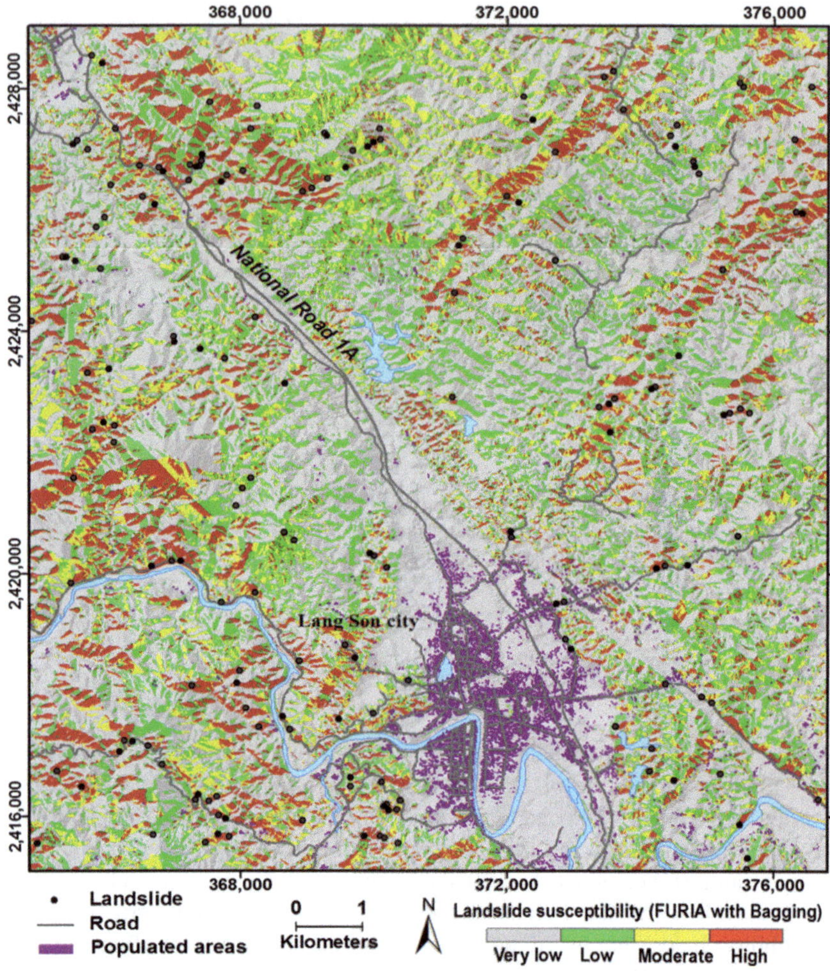

Fig. 5 Landslide susceptibility map of the Lang Son city areas using the FURIA with bagging model

4 Validation and Comparison of Landslide Susceptibility Models

4.1 Model Performance and Evaluation

The performance measurement of four landslide susceptibility models (FURIA, FURIA with Bagging, Decision tree, Decision tree with Bagging) were assessed using several statistical evaluation criteria (Tien Bui et al. 2012a) as follows:

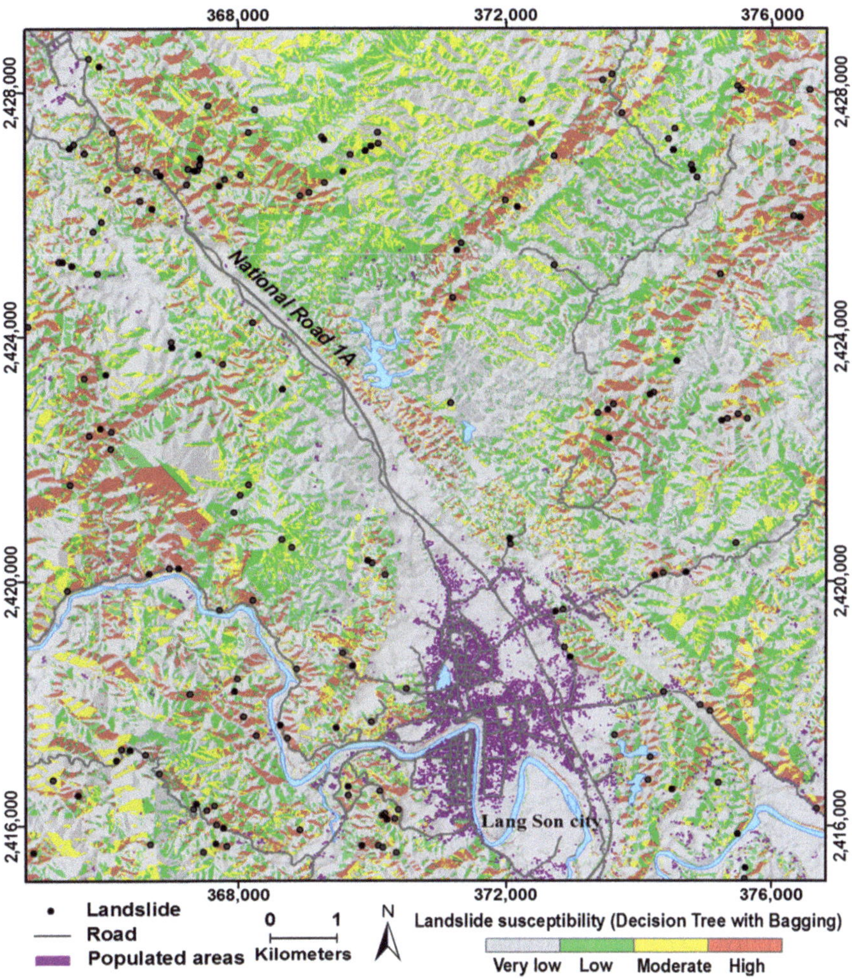

Fig. 6 Landslide susceptibility map of the Lang Son city area using the decision tree with bagging model

$$Sensitivity = TP/(TP + FN) \tag{9}$$

$$Specificity = TN/(TN + FP) \tag{10}$$

$$Accuracy = (TP + TN)/(TP + TN + FN + FP) \tag{11}$$

$$F - measure = 2 * Sensitivity * Specificity/(Sensitivity + Specificity) \tag{12}$$

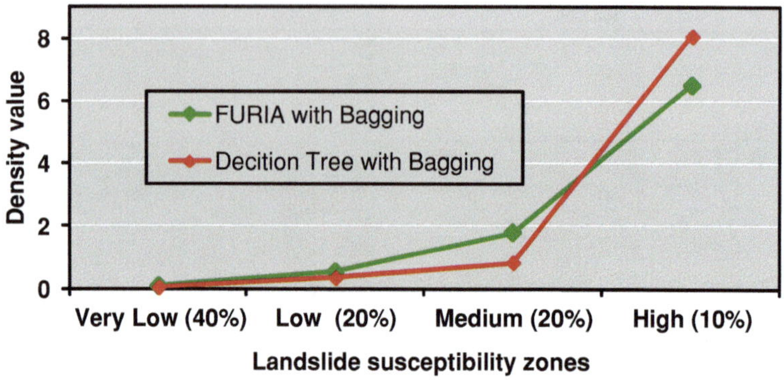

Fig. 7 Density plots of four landslide susceptibility classes of the FURIA with bagging and the decision tree with bagging models

$$Root\,mean\,squared\,error\,(RMSE) = \mathrm{Sqrt}\left(\frac{(pred_1 - act_1)^2 + \cdots + (pred_n - act_n)^2}{n}\right) \tag{13}$$

$$Mean\,absolute\,error\,(MAE) = \frac{|pred_1 - act_1| + \cdots + |pred_n - act_n|}{n} \tag{14}$$

$$\begin{aligned} &Kappa\,index\,(\kappa) = \frac{P_C - P_{exp}}{1 - P_{exp}} \\ &\quad \text{where}\, P_C = (TP + TN)/(TP + TN + FN + FP) \\ &\qquad P_{exp} = [(TP + FN)(TP + FP) + (FP + TN)(FN + TN)]/Sqrt(TP + TN + FN + FP) \end{aligned} \tag{15}$$

True positive (*TP*) rate measures the proportion of number of pixels that are correctly classified as landslides. True negative (*TN*) rate measures the proportion of number of pixels that are correctly classified as non-landslide. False negatives (*FN*) are the number of landslide pixels classified as non-landslide pixel. True negatives (*FN*) are the number of non-landslide pixels classified as landslide pixels. Precision measures the proportion of the number of pixels that are correctly classified as landslide occurrences. *F-measure* combines precision and sensitivity into their harmonic mean. *Act* is the actual target value whereas *pred* is the predicted value. P_C is the proportion of number of pixels that are correctly classified as landslide or non-landslide. P_{exp} is the expected agreements.

It could be observed that there is a high and almost equal in term of classification accuracy for the three models, FURIA with Bagging, the Decision Tree, and the Decision Tree with Bagging (Table 4). Accuracy assessment by classes (Table 5) shows that the rate of correctly classified landslide pixels is higher than those for non-landslide pixels for all models.

Table 6 The range of the kappa index and the corresponding agreement between the model and reality (Cohen 1960)

No.	Kappa index	Agreement
1	0.80–1	Almost perfect
2	0.60–0.80	Substantial
3	0.40–0.60	Moderate
4	0.20–0.40	Fair
5	0–0.2	Slight
6	≤0	Poor

The reliability of the susceptibility models was measured using the Kappa index (Guzzetti et al. 2006). Kappa indexes for the FURIA, FURIA with Bagging, the Decision Tree, and the Decision Tree with Bagging are 0.697, 0.728, 0.736, and 0.750 respectively. It indicates a substantial agreement (Table 6) between the observed and the predicted values. The reliability analysis results are satisfying compared with other works such as Saito et al. (2009) and Tien Bui et al. (2012a).

4.2 Model Validation

The prediction capability of the susceptibility models were evaluated using ROC curves. A ROC curve is used to plot sensitivity/1-specificity with different thresholds. Comparied to the to success and prediction rate curves (Chung and Fabbri 2003), ROC curves are considered not sensitive, by keeping in mind of the considerable difference between landslide and non-landslide pixels. Therefore ROC curves are considered as more appropriate evaluation and validation tool for landslide models (Van Den Eeckhaut et al. 2009).

The area under the ROC curve (AUC) is used as an important measurement of the landslide model performance. A landslide model will be considered a preferred model if it has a larger AUC value than other models. A perfect model will have an AUC of 1 whereas a random model has an AUC of approximately 0.5.

In this study, ROC curves and AUCs were prepared for each landslide model in two cases: the first one used the training dataset and the second one used the validation dataset. Since in the first case the same landslide pixels that have already been used to construct the landslide models, therefore, the ROC curve and AUC is only measured the degree of model fit of the model with the training dataset. The result (Fig. 8 and Table 7) shows that the highest degree of fit has the Decision Tree with Bagging (AUC = 0.925), followed by the FURIA with Bagging (AUC = 0.913), the Decision Tree (AUC = 0.908), and FURIA (AUC = 0.878). The prediction capabilities of the landslide models were obtained in the second case. This case uses the validation dataset that has not been used in the training phase and can provide the validation and explain how well the model and the conditioning factors predict the existing landslides (Pradhan and Lee 2010c). The result (Fig. 9 and Table 8) shows that the FURIA with Bagging has the highest prediction capability (AUC = 0.802). The remaining models have almost equal prediction capability (AUC from 0.773 to 0.783).

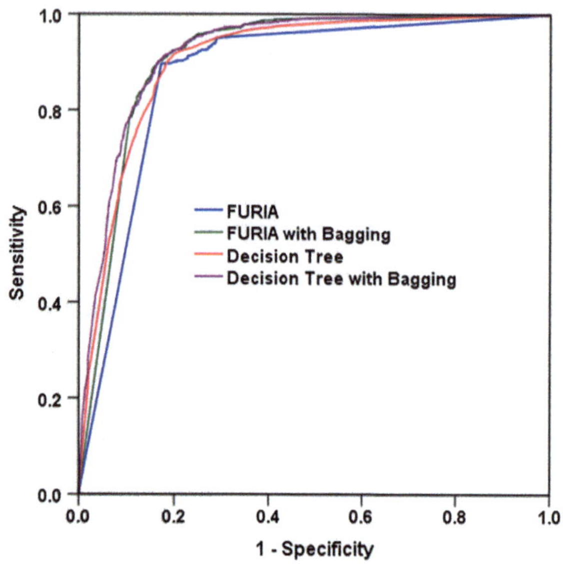

Fig. 8 ROC curves based on the training dataset for the FURIA, the FURIA with bagging, the decision tree, and the decision tree with bagging

Table 7 Area under the curves (*AUC*) based on the training dataset for the FURIA, the FURIA with bagging, the decision tree, and the decision tree with bagging

No.	Landslide model	AUC	Std. error	95 % CI	
				Lower bound	Upper bound
1	FURIA	0.878	0.004	0.869	0.886
2	FURIA with bagging	0.913	0.004	0.906	0.920
3	Decision tree	0.908	0.004	0.901	0.915
4	Decision tree with bagging	0.925	0.003	0.919	0.931

4.3 Relative Contribution of the Conditioning Factors

The relative contribution of each conditioning factor on the susceptibility models can be estimated by excluding the factor in the models and then the classification accuracy was estimated. It is clear that the highest accuracy was obtained when all of the six factors are used (Table 9). Aspect has the highest contribution to the models whereas soil type has the lowest contribution. More details are shown in Table 9.

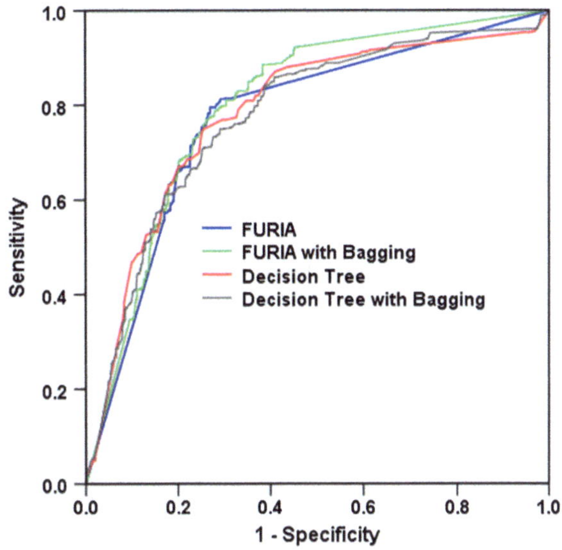

Fig. 9 ROC curves based on the validation dataset for the FURIA, the FURIA with bagging, the decision tree, and the decision tree with bagging

Table 8 Area under the curves (*AUC*) based on the validation dataset for the FURIA, the FURIA with bagging, the decision tree, and the decision tree with bagging

No.	Landslide model	AUC	Std. error	95 % CI	
				Lower bound	Upper bound
1	FURIA	0.773	0.008	0.757	0.790
2	FURIA with bagging	0.802	0.008	0.786	0.817
3	Decision tree	0.783	0.008	0.767	0.799
4	Decision tree with bagging	0.777	0.008	0.761	0.793

Table 9 Relative contribution of the conditioning factors

No	Conditioning factors	Classification accuracy (%)	
		FURIA with bagging	Decision tree with bagging
1	Minus slope	82.61	83.54
2	Minus aspect	79.08	79.54
3	Minus lithology	83.39	83.60
4	Minus landuse	82.64	83.68
5	Minus soil type	85.61	85.75
6	Minus distance to faults	83.07	83.10
7	All	86.38	87.50

5 Conclusion

Over the last two decades, various methods and techniques for the landslide modeling have been used and discussed, however, the FURIA model and Bagging technique are seldom been applied and a comparison between FURIA with

Decision Tree and their Bagging has not been carried out so far. Decision Tree models have only been applied in a limited number of studies. The recent development in geographic information systems (GIS) and computer science allows users to apply these techniques with huge GIS data (Pradhan 2013).

In general, there are three main steps used for the landslide susceptibility modeling in this study, data preparation, susceptibility analyses, and validation and comparison. In the first step, the landslide inventory map with 172 landslide polygons was constructed. Among them, approximately 70 % (117 cases) was selected for the training models, whereas the remaining 30 % (55 cases) were used for model validation. And then, landslide conditioning factors were determined. All maps were prepared with a spatial resolution of 5 m. In the next step, a total of four models were constructed. The validation result show that the FURIA with Bagging (AUC = 0.913) and the Decision Tree with Bagging (0.925) have the highest degree of fit with the training data. They are followed by the Decision Tree (AUC = 0.908), and FURIA (AUC = 0.878). Regarding the prediction capability, the FURIA with Bagging has the highest value (AUC = 0.802), the other models have almost equal prediction capability (AUC is around 0.77).

It is well known that the selection of sampling strategy influences the prediction capability of landslide models (Yilmaz 2010). As shown in Chung and Fabbri (2008), the temporal partitioning of landslides is considered to be the best method. However, the temporal partitioning method is not suitable for this study due to unknown dates of landslide occurrence. Therefore the randomly split method was used. The main disadvantage of this method is that it may cause an overestimated of prediction capability of future landslides if spatial separation between training and validation landslides are small (Brenning 2005).

The selection of conditioning factors are an important task for the assessment of landslide susceptibility and may impact on the overall prediction performance for landslide susceptibility models (Pradhan 2013). Although no agreement on universal guidelines has been reached for the selection of conditioning factors (Tien Bui et al. 2012b), the factors related to topography, geology, soil types, hydrology, geomorphology, and land use are considered to be the most commonly used in landslide analyses (Van Westen et al. 2008). Therefore six landslide conditioning factors (slope, aspect, lithology, distance to faults, landuse, and soil type) were selected for this study.

As a final conclusion, all the models exhibit reasonably satisfactory performance. However, we may conclude that the FURIA with Bagging is considered to be the best one from this study. And it is important to note that the performance of these landslide models depends not only on the methods but also on sampling strategy followed, as well as the quality of the data used. Therefore, the quality of the susceptibility maps produced by the four models can be improved if the quality of the data used increases. The analyzed result obtained from this study is valid for shallow landslides. These maps may be useful for natural hazard management policy, planning and decision-making in landslide prone areas.

Acknowledgement This research was supported by the Geomatics Section, Department of Mathematical Sciences and Technology, Norwegian University of Life Sciences, Norway.

References

Akgun A, Sezer EA, Nefeslioglu HA, Gokceoglu C, Pradhan, B (2012) An easy-to-use MATLAB program (MamLand) for the assessment of landslide susceptibility using a Mamdani fuzzy algorithm. Comput Geosci 38:23-34

Althuwaynee OF, Pradhan B, Park HJ, Lee JH (2014) A novel ensemble decision-tree based CHi-squared automatic interaction detection (CHAID) and multivariate logistic regression models in landslide susceptibility mapping. Landslides (Article online first available). http://dx.doi.org/10.1007/s10346-014-0466-0

Breiman L (1996) Bagging predictors. Mach Learn 24:123–140

Breiman L, Friedman JH, Olshen RA, Stone CJ (1984) Classification and regression trees. Wadsworth, Belmont

Brenning A (2005) Spatial prediction models for landslide hazards: review, comparison and evaluation. Nat Hazards Earth Syst Sci 5:853–862

Chacon J, Irigaray C, Fernandez T, El Hamdouni R (2006) Engineering geology maps: landslides and geographical information systems. Bull Eng Geol Environ 65:341–411

Chung C-J, Fabbri AG (2008) Predicting landslides for risk analysis—spatial models tested by a cross-validation technique. Geomorphology 94:438–452

Chung CJF, Fabbri AG (2003) Validation of spatial prediction models for landslide hazard mapping. Nat Hazards 30:451–472

Cohen J (1960) A coefficient of agreement for nominal scales. Educ Psychol Measur 20:37–46

Cohen WW (1995) Fast effective rule induction. In: Machine learning: proceedings of the twelfth international conference. Morgan Kaufmann, Lake Taho

Glade T, Anderson M, Crozier MJ (2005) Landslide hazard and risk. Wiley, London

Guzzetti F, Carrara A, Cardinali M, Reichenbach P (1999) Landslide hazard evaluation: a review of current techniques and their application in a multi-scale study, Central Italy. Geomorphology 31:181–216

Guzzetti F, Reichenbach P, Ardizzone F, Cardinali M, Galli M (2006) Estimating the quality of landslide susceptibility models. Geomorphology 81:166–184

Huhn J, Hullermeier E (2010) An analysis of the FURIA algorithm for fuzzy rule induction. In: Koronacki J, Raś Z, Wierzchoń S, Kacprzyk J (eds) Advances in machine learning I, vol 262. Springer, Berlin, pp 321–344

Huhn J, Hullermeier E (2009) FURIA: an algorithm for unordered fuzzy rule induction. Data Min Knowl Disc 19:293–319

Kanungo D, Arora M, Gupta R, Sarkar S (2008) Landslide risk assessment using concepts of danger pixels and fuzzy set theory in Darjeeling Himalayas. Landslides 5:407–416

Korup O, Gorum T, Hayakawa Y (2012) Without power? Landslide inventories in the face of climate change. Earth Surf Proc Land 37:92–99

Lim TS, Loh WY, Shih YS (2000) A comparison of prediction accuracy, complexity, and training time of thirty-three old and new classification algorithms. Mach Learn 40:203–228

Michael JA, Gordon SL (1997) Data mining technique: for marketing, sales and customer support. Wiley, New York

Pourghasemi H, Pradhan B, Gokceoglu C (2012) Application of fuzzy logic and analytical hierarchy process (AHP) to landslide susceptibility mapping at Haraz watershed. Iran Nat Hazards 63:965–996

Pradhan B (2013) A comparative study on the predictive ability of the decision tree, support vector machine and neuro-fuzzy models in landslide susceptibility mapping using GIS. Comput Geosci 51:350-365

Pradhan B, Lee S (2010a) Delineation of landslide hazard areas on Penang Island, Malaysia, by using frequency ratio, logistic regression, and artificial neural network models. Environ Earth Sci 60:1037–1054

Pradhan B, Lee S (2010b) Landslide susceptibility assessment and factor effect analysis: backpropagation artificial neural networks and their comparison with frequency ratio and bivariate logistic regression modelling. Environ Model Softw 25:747–759

Pradhan B, Lee S (2010c) Regional landslide susceptibility analysis using back-propagation neural network model at Cameron Highland, Malaysia. Landslides 7:13–30

Pradhan B, Sezer EA, Gokceoglu C, Buchroithner MF (2010) Landslide susceptibility mapping by neuro-fuzzy approach in a landslide-prone area (Cameron Highlands, Malaysia). IEEE Trans Geosci Remote Sens 48:4164–4177

Quinlan JR (1993) C4.5 programs for machine learning. Morgan Kaufmann, San Mateo

Quinlan JR (1986) Induction of decision trees. Mach Learn 1:81–106

Quinlan JR, Cameron-Jones RM (1993) FOIL: a midterm report. In: European conference on machine learning. Springer, Berlin

Quoc NK, Dan TH, Hung L, Huyen DT (1992) Geological map. In: Binh Gia group (ed), Vietnam Institute of Geosciences and Mineral Resources, Hanoi

Roberds W (2005) Estimating temporal and spatial variability and vulnerability. In: Hungr O, Fell R, Couture R, Eberhardt E (eds) Landslide risk management. Taylor and Francis, London

Rokach L (2010) Ensemble-based classifiers. Artif Intell Rev 33:1–39

Saito H, Nakayama D, Matsuyama H (2009) Comparison of landslide susceptibility based on a decision-tree model and actual landslide occurrence: The Akaishi Mountains, Japan. Geomorphology 109:108–121

Sassa K, Canuti P (2008) Landslides-disaster risk reduction. Springer, Berlin, p 650

Sezer EA, Pradhan B, Gokceoglu C (2011) Manifestation of an adaptive neuro-fuzzy model on landslide susceptibility mapping: Klang valley, Malaysia. Expert Syst Appl 38:8208–8219

Tam VT, Tuy PK, Nam NX, Tuan LC, Tuan ND, Trung ND et al (2006) Geohazard investigation in some key areas of the northern mountainous area of Vietnam for the planning of socio-economic development. Vietnam Institute of Geosciences and Mineral Resources, Hanoi, p 83

Tehrany MS, Pradhan B, Jebur MN (2013) Spatial prediction of flood susceptible areas using rule based decision tree (DT) and ensemble bivariate and multivariate statistical models. J Hydrol 504:69-79. http://dx.doi.org/10.1016/j.jhydrol.2013.09.034

Tien Bui D, Pradhan B, Lofman O, Revhaug I (2012a) Landslide susceptibility assessment in Vietnam using support vector machines, Decision tree and Naïve Bayes models. Math Prob Eng. doi.10.1155/2012/974638

Tien Bui D, Pradhan B, Lofman O, Revhaug I, Dick O (2013a) Regional prediction of landslide hazard using probability analysis of intense rainfall in the Hoa Binh province, Vietnam. Nat Hazards 2:707–730

Tien Bui D, Ho TC, Revhaug I, Pradhan B, Nguyen DB (2013b) "Landslide Susceptibility Mapping Along the National Road 32 of Vietnam Using GIS-Based J48 Decision Tree Classifier and Its Ensembles." In Cartography from Pole to Pole, edited by Buchroithner M, Prechtel N, Burghardt D, 303-17. Springer Berlin Heidelberg

Tien Bui D, Pradhan B, Lofman O, Revhaug I, Dick OB (2012b) Landslide susceptibility assessment in the Hoa Binh Province of Vietnam: a comparison of the Levenberg–Marquardt and Bayesian regularized neural networks. Geomorphology 171–172:12–29

Tien Bui D, Pradhan B, Lofman O, Revhaug I, Dick OB (2012c) Landslide susceptibility mapping at Hoa Binh province (Vietnam) using an adaptive neuro-fuzzy inference system and GIS. Comput Geosci 45:199–211

Trawinski K, Cordon O, Quirin A (2011) On designing fuzzy rule-based multiclassification systems by combining furia with bagging and feature selection. Int J Uncertainty Fuzziness Knowl Based Syst 19:589–633

Truong PD, Nghi TH, Phuc PN, Quyet HB, The NV (2009) Geological mapping and mineral resource investigation at 1:50 000 scale for Lang Son area. Northern Geological Mapping Division, Hanoi

Van Den Eeckhaut M, Reichenbach P, Guzzetti F, Rossi M, Poesen J (2009) Combined landslide inventory and susceptibility assessment based on different mapping units: an example from the Flemish Ardennes, Belgium. Nat Hazards Earth Syst Sci 9:507–521
Van Westen CJ, Castellanos E, Kuriakose SL (2008) Spatial data for landslide susceptibility, hazard, and vulnerability assessment: an overview. Eng Geol 102:112–131
Witten IH, Frank E (2005) Data mining: practical machine learning tools and techniques, 2nd edn. Morgan Kaufmann, Los Altos
Yilmaz I (2010) The effect of the sampling strategies on the landslide susceptibility mapping by conditional probability and artificial neural networks. Environ Earth Sci 60:505–519
Zhao Y, Zhang Y (2008) Comparison of decision tree methods for finding active objects. Adv Space Res 41:1955–1959

Application of Geo-Spatial Technique for Flood Inundation Mapping of Low Lying Areas

Dhruvesh P. Patel and Prashant K. Srivastava

Abstract Flooding is one of the severe disaster causes mass demolition of properties and affected human lives. In hazardous flood of year 2006, 90–95 % of Surat city, India, was under water and so that local planner as well as decision makers need accurate information on the spatial distribution, magnitude, depth of flooding and land use affected by such floods. Surat city is majorly partitioned into seven different zones named north zone, east zone, central zone, south zone, south-east zone, south-west zone, and west zone. Purpose of this study is to determine inundation of water in low laying areas of west zone. By procedure of Geo-reference along with Ground Control Point (GCP) and GPS points, 0.5 m interval contour map for west zone is introduced. Digitization of contour through GIS software and Digital Elevation Model (DEM) of West Zone through ArcGIS software is carried out. Probable submergence area for rescue work is also scrutinized. Graph of submergence area of West Zone according Town Planning Scheme (TPS) versus water level and flood Inundation map are generated which specify that West Zone and its TPS are low lying areas in Surat whose 20–25 km^2 area will be submerge when water level exceeds 12 m height (MSL). The accuracy and validation of DEM is calculated by comparison with actual observed data at the time of flooding.

Keywords Digital elevation model · Flood · Inundation mapping · RS & GIS · Vulnerability

D. P. Patel (✉)
Department of Civil Engineering, Saffrony Institute of Technology,
S.P.B. Patel Engineering College, Linch, Mehsana, Gujarat 384435, India
e-mail: dhruvesh1301@gmail.com

P. K. Srivastava
Department of Civil Engineering, University of Bristol, Bristol BS81TR, UK

P. K. Srivastava et al. (eds.), *Remote Sensing Applications in Environmental Research*,
Society of Earth Scientists Series, DOI: 10.1007/978-3-319-05906-8_7,

1 Introduction

Flood is defined as extremely high flow of water from river, lake, pond, reservoir or any other water body, wherein water inundates outside of the water body area. Flood may also occurs when the sea level rises extremely or above the coastal lands due to tidal sea and sea surges. The flooding of homes, industrial areas, and amenities in recent years, along with an appreciation of the likely consequences of climate change, has brought the issue of flooding to the attention of the media and the general public (Wright et al. 2008). On the global scale, storms and floods are the most destructive of natural disaster and cause the greatest number of deaths and property loss (Casale and Margottini 1999; Sadrolashrafi et al. 2008). Flooding and flash flooding poses serious hazards to human populations in many parts of the world. Surat, the study area, is one of the cities of state Gujarat, India and has been frequently affected by flood in river Tapi. Due to augmented industrialization in and around Surat, the population is reached up to 50.64 lakh in year 2011 which was 28.76 lakh in year 2001. Surat has experienced three major flood events in recent past in year 1994, 1998 and 2006 which caused heavy loss of property and human lives. In year 2006, large amount of water spilled from low rise river sections resulted into flood and 90–95 % Surat city was under submerge up to 4–5 m (Patel and Dholakia 2010; Joshi et al. 2012; Patel et al. 2012b). This unprecedented flood in Surat caused damages of over Rs. 21,000 crores approximately.

High resolution, high accuracy topographic data sets are becoming increasingly available for flood prediction studies in a number of countries (Bates et al. 2003). Remote Sensing, Geo graphical Information System and Global Positioning System has been demonstrated to be a useful tool in mapping flood extent (Horritt and Bates 2001; Horritt et al. 2001) and hence for validating numerical inundation models (Horritt and Bates 2002). Various methodologies or approaches have been adopted by researchers in relevant flood research in the period 1997–2010 (Islam and Sado 2000; Kresch et al. 2002; Pistocchi and Mazzoli 2002; Sutanta 2002; Tralli et al. 2005; Wright et al. 2008; Andrysiak and Maidment 2000). Flood inundation models commonly require four key data items: (1) topographic data to construct the model grid; (2) bulk flow or stage data to provide model inflow and outflow boundary conditions; (3) an estimate of the effective friction parameter for each model cell; and (4) a source of validation data (Wright et al. 2008). The elevation information in a digital format using image matrix in which the value of each pixel is associated with a specific topographic height is called DEM, is a computerized representation of the Earths relief in 3D (Patel and Dholakia 2010; Patel et al. 2012a, b). Triangulated Irregular Networks (TIN), regular grids, contour lines and scattered data points (Sulebak 2000) are some methods usually used for generating DEM.

Using DEM as an input, flood risk map, flood hazards map, geomorphologic map, Vulnerability analysis, damage estimation and risk assessment, hydraulic modeling, flood inundation and storm-tide flood-inundation maps can be generated

(Islam and Sado 2000; Kresch et al. 2002; Pistocchi and Mazzoli 2002; Sutanta 2002; Tralli et al. 2005; Wright et al. 2008; Andrysiak and Maidment 2000; Patel and Dholakia 2010). Two-dimensional flood inundation models are widely used tools for flood hazard mapping and an essential component of statutory flood risk management guidelines in many countries (Neal et al. 2012). Noman et al. (2001, 2003), Agnihotri and Patel (2008), Patel et al. (2012a) also described the two primary inputs to any automated floodplain delineation process in 3D, which are creation of a water surface using Geo-referenced water level points and DEM. The flood depth map is obtained by subtracting the DEM from the water surface.

Generation of DEM, flood inundation map and submerge area of West Zone (Stage-area submerge curve) are prepared and also verified in this study.

2 Study Area and Datasets

2.1 *Tapi Basin*

Tapi is the second largest westward draining inter-state river after mighty Narmada, having its source at Multai, Betul district, Madhya Pradesh, India. The Tapi basin is the northern most basin of the Deccan plateau and is situated between North latitudes 20–22°N and East longitude 72–78°E approximately (CWC 2000–2001). Tapi basin is basically divided into three zones, viz. Upper Tapi Basin (UTB), Middle Tapi Basin (MTB) and Lower Tapi Basin (LTB). The river has a total length of 724 km, out of which the last lap of 214 km is in Gujarat state and it meets an Arabian Sea in the Gulf of Cambay approximately at 19.2 km west of Surat city. Tapi covers an area of, approximately, 51,504 km^2 (79 %) in Maharashtra, 9,804 km^2 (15 %) in Madhya Pradesh and 3,837 km^2 (6 %) in Gujarat state.

2.2 *Surat City*

Surat is situated between latitude 21°06′N–21°15′N and longitude 72°45′E–72°54′E, on the bank of river Tapi and having coastline of Arabian Sea on its west. Surat falls in Survey of India map number 46C/15, 16. Tapi River and Surat city with zone boundary is shown in Fig. 1. Surat city lies at a bend of the river Tapi, where its course swerves suddenly from the south-east to south-west. From the right bank of the river, ground rises slightly towards the north but the height above Mean Sea Level (MSL) is 13 m. The topography is controlled by the river which is flat in general but having gentle slope from north-east to south-west.

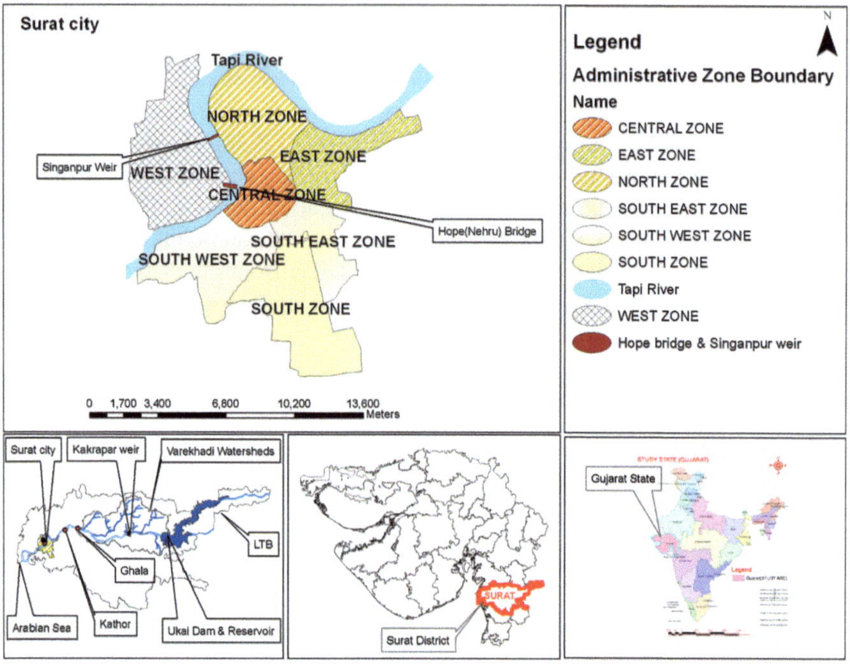

Fig. 1 Location map of LTB, Ukai dam, Varekhadi watershed, Kakrapar weir, Ghala, Kathor, Singanpur weir, Hope (Nehru) bridge, Surat city with administrative zone boundary

2.2.1 Geology and Soil Conditions

Study area can be divided into two geomorphic units namely, coastal and alluvial. The coastal area represents marshy shoreline with an extensive tidal flat stretch intercepted by estuaries. Alluvial deposited by Tapi has formed the alluvial area which is covered by recent alluvium of quaternary age. The alluvial plain is characterized by flood plain of the Tapi and Mindhola River where there is a thick alluvial cover. The alluvial plain merges into a dry, barren, sandy coastal area. The coastal area around the river is covered by mud. The alluvium consists of sand and clay layers.

2.2.2 Ground Water Table

Pre and post monsoon water levels of different open wells in and around Surat during year 1970–2006 indicated that an average rise in water level is about 1.85 m and average standard deviation of rise in water level is about 0.93 m. The ground water level generally rises 2–5 m during monsoon period (June–October) while during the rest of the year, it drops down to 5 m and even up to 10 m at some locations.

2.2.3 Climate

The climate of Surat can be broadly divided into four seasons namely summer, rainy, autumn and winter. Duration of summer season is of 3 months, from March to May, rainy from June to September, autumn in October–November and winter from December to February.

2.2.4 Temperature and Rainfall

Summer season in study area is quite hot with temperature ranges from 37.78 to 44.44 °C. The climate is pleasant during the monsoon while in autumn it is temperate. Winter is not so cold but the temperature in January month ranges from 10 to 15.5 °C. The average annual rainfall of the city is 1,143 mm.

2.2.5 Demography/Population in the Study Area

With the beginning of 20th century, Surat had started developing its sub-urban areas namely Udhna, Athwa, Fulpada etc. along with the various corridors opened up through various gates in the radial pattern. The city was originally established on the south-eastern bank of the river Tapi with a castle at the eastern bank and a custom house at the northern side of the castle. In the beginning, the activities were concentrated within the inner wall which had been constructed in year 1664. At that time, the area of the city within this inner wall was 178 hectors. The evolution of the city from 1494 to 2004 AD is shown in Fig. 2.

2.3 Hydraulic Structures at LTB

The LTB contains hydraulic structures namely Ukai dam, Kakrapar weir and Singanpur weir. Details of these structures are as under.

2.3.1 Ukai Dam

The Ukai dam is located about 30 km U/S of the Kakrapar weir and about 95 km U/S of Surat city, originally approved by the planning commission of government of India in 1969 and the construction of the dam was completed in 1973 (Fig. 3). The Ukai reservoir at its FRL of 105.15 m (345 ft.) has a live storage capacity of 7,369 Mm^3 with water spread of about 600 km^2 and maximum length of about 112 km. The reservoir is expected to attain Maximum Water Level (MWL) of 106.99 m (351 ft.) while passing the Probable Maximum Flood (PMF) of 59,747 m^3/s (21.16 lakh cusecs). Ukai dam is a composite dam with a maximum

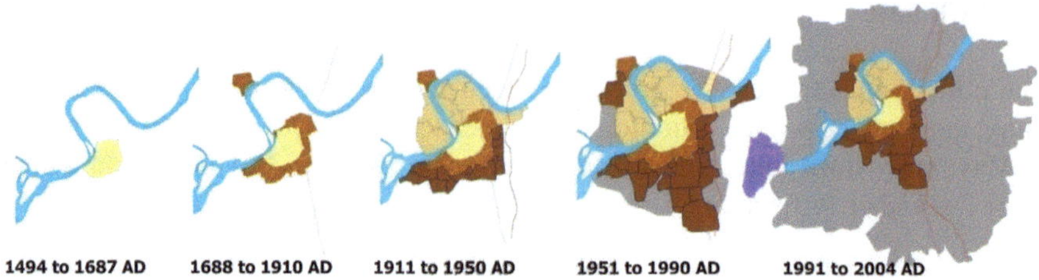

Fig. 2 Evolution of Surat city from year 1494 to 2004 (*Source* SMC, vision 2020)

Fig. 3 Ukai Dam

height of about 80.7 m above its deepest foundation level. The total length of dam is 4,927 m, out of which 4,058 m is earthen dam of zoned fill type. Masonry gravity dam, including the 425 m long spillway and the power dam occupies the remaining length. The spillway, located in the left bank of the river, is provided with 22 no. of radial crest gates of 15.55 m × 14.78 m (51 ft. × 48.5 ft.) size. The maximum discharge capacity of the spillway at its MWL is 46,270 m^3/s (16.34 lakh cusecs) while the corresponding discharging capacity of the spillway at FRL is 37,860 m^3/s (13.37 lakh cusecs). The power house, located at the toe of the dam on the left side of the spillway, is equipped with four units of 75 MW each and generally operates as a peaking station. The water released from the dam to powerhouse is picked up at the Kakrapar weir in downstream for firming up the irrigation in the Kakrapar canal system. For the purpose of flood protection and moderation, a reservoir operation schedule is prepared by CWC, New Delhi, for the monsoon period. The overall behavior of Ukai dam in last 34 years during the flood in river Tapi, is shown in Fig. 4. The massive flood years are 1959, 1968, 1978, 1979, 1994, 1998 and 2006. Ukai reservoir provides protection against heavy floods to an area of 827 km^2 on the downstream (D/S). Later, by constructing the flood embankments on both the banks of the river Tapi, between Kathor and Surat, it provides protection to an additional area of 230 km^2. Thus, the Ukai reservoir and the flood embankment, together provide protection to

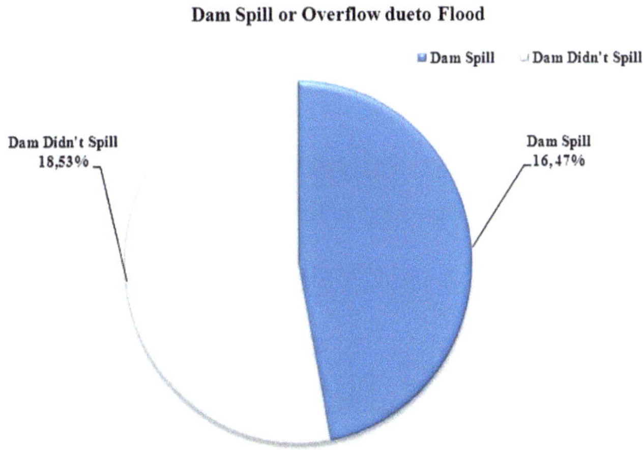

Fig. 4 Behavior of Ukai Dam

1,057 km^2 of land area, saving approximately 2 million inhabitants residing in greater Surat region around river banks. Flood embankments are designed for uniform river discharge of 24,069 m^3/s (8.5 lakh cusecs). The Safe carrying capacity of river at Surat is estimated to be 11,326 m^3/s (4.0 lakh cusecs).

2.3.2 Kakrapar Weir

The scheme comprises of an Ogee shaped masonry pick up weir constructed across the Tapi River near Kakrapar, Surat district, Gujarat. Kakarapar weir (Fig. 5) is situated 30 km away from Ukai dam at D/S and is built for irrigation purpose in nearby area. It is 621 m long with 14 m height and having catchment area about 62,801 km^2. It leads a maximum discharge of 38,228 m^3/s from its Ogee type of spillway. The weir is having 442 km^2 reservoir area, where the gross storage capacity is 51.51 Mm^3 and the effective storage capacity is 36.51 Mm^3. Two canals take off from right and left bank of weir by canal head regulating work. Right canal is 219.74 km (including main canal, branches canal, minor and major distributaries) long and having the discharge caring capacity 70.23 m^3/s (main canal). It covers the gross command area of 100,220 × 10^4 m^2 and culturable command area of 58,745 × 10^4 m^2. Same way, the left canal is 296.91 km (including main canal, branches canal, minor and major distributaries) long and having the discharge caring capacity 85.63 m^3/s (main canal). It covers the gross command area 247,000 × 10^4 m^2 and culturable command area 145,335 × 10^4 m^2. This project was commissioned in the year 1954 as stage-I of the Ukai project-3.

Fig. 5 Kakrapar Weir

Fig. 6 Singanpur Weir

2.3.3 Singanpur Weir

Singanpur causeway (Fig. 6) cum weir was constructed in 1995 with cost of Rs. 31 crores on river Tapi near Singanpur–Rander village. The length of bridge is 580 m, out of which, 98 m is gated portion and the remaining is ungated. The bridge has 16 span and length of each span is 6.20 m. Length of retaining walls is 80 and 100 m for left and upper side respectively. The design discharge of the weir was estimated as 680 m^3/s (24,000 cusecs) with designed HFL of 16 m in U/S and 14 m in D/S. The foundation of the bridge is open integrated and substructure with reinforced cement concrete abutments and piers.

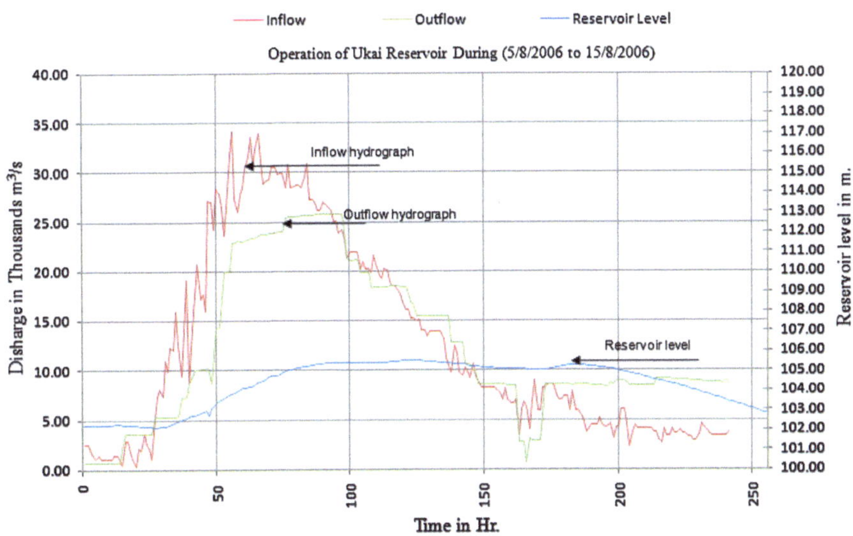

Fig. 7 Operation (inflow/outflow) of Ukai reservoir, 2006

2.4 Flood Event 2006

2006 flood had created disaster in Surat city and surrounding areas which falls in semi-arid and arid agro-climate region. Flood event during 7–14th August in year 2006, resulted in mass devastation of lives and belongings damage worth INR 21,000 crores in Surat and Hazira twin city. For several weeks, whole society was lifeless and many diseases were reported in city (Patel and Srivastava 2013). Operation of Ukai reservoir during 5 days of August, 2006 is shown in Fig. 7. As indicated in graph, the water level rise continuously from 102.108 m (335 ft.) to 105.344 m (345.62 ft.) as maximum, after 4 days. The Ukai reservoir was having large catchment area and due to constant inflow, the large volume of water accumulated in it. The graph shown indicates that on 7th August, compare to inflow, water released from dam was very less, almost half. The dam operation during 10 days of August 2006, combining inflow–outflow and reservoir level, is shown by hydrograph. The inflow in the dam was continuously fluctuating for first 20 h from 05/08/2006 and then suddenly rose up to 16,423 m^3/s (5.8 lakh cusecs), whereas the outflow from the dam was below 5,663 m^3/s (2 lakh cusecs) for that period. Next 36 h, from 07/08/2006 afternoon, were very critical and during that period the inflow touched to 33,980 m^3/s (12 lakh cusecs) which is the maximum in the recent flood history and to manage the situation, the water released from the dam was about 25,768 m^3/s (9.1 lakh cusecs) for 24 h (Table 1). The inflow was then down to 12,742.42 m^3/s (4.5 lakh cusecs) and out flow was maintained almost same.

Table 1 Flood Event, 2006

Date	Time	In-flow (m^3/s)	Out-flow (m^3/s)	Reservoir level (m)	Level at Nehru Bridge (m)
5-8-2006	8.00 a.m.	2,434.03	670.08	102.19	6.50
	8.00 p.m.	1,374.88	669.40	102.26	7.40
6-8-2006	8.00 a.m.	2,126.20	3,537.29	102.14	8.40
	8.00 p.m.	9,350.56	7,214.47	102.56	9.14
7-8-2006	8.00 a.m.	24,172.10	11,581.56	103.46	9.88
	8.00 p.m.	30,374.50	23,107.28	104.21	11.70
8-8-2006	8.00 a.m.	29,821.20	23,901.73	104.97	12.35
	8.00 p.m.	27,225.40	25,768.01	105.32	12.85
9-8-2006	8.00 a.m.	24,238.90	24,069.02	105.34	12.90
	8.00 p.m.	20,149.10	18,405.72	105.38	10.80

2.5 Data Collection

For this research work, Geo-coded Indian Remote Sensing (IRS-1D) satellite image of the study area for April 2005 from BISAG, Gandhinagar is used. Topographical sheets with the scale of 1:50,000 are collected from SOI, Ahmedabad. High resolution Google-earth images (http://earth.google.com) of the study area are downloaded for thematic maps creation. Physical measurements for the river hydraulic parameters after monsoon 2006 is collected from Surat Municipal Corporation (SMC) while the contour maps for various zones of the city at 0.5 m interval are generated using ArcGIS 9.2 software from the topographical sheets. Trimble Geo explorer XT global positioning system is used to survey the West Zone area of Surat City. In T.P. roads of West Zone and on left-right embankments of River Tapi, field survey has been performed. The water level and river discharge data from hourly to daily scales at Nehru Bridge are collected from CWC, SWDC and Irrigation department. Zonal map of the city is collected from SMC.

3 Methodology

At an early stage, ERDAS 9.1 software, Geo-referenced by ArcGIS 9.2, is used to combined image pieces of study area taken from Google-Earth. Counter map is then Geo-referenced over downloaded images (Fig. 8) and after that spatial adjustment is applied for precision. Digitized counters of 0.5 m interval through GIS software (Fig. 9) and GPS point data are integrated (Fig. 10). Temporal and spatial data are linked using GIS as well as customized Data Base Management System (DBMS). Digitized counters are then converted into TIN model using 3D analyst tool which is finally converted into DEM using spline interpolation method

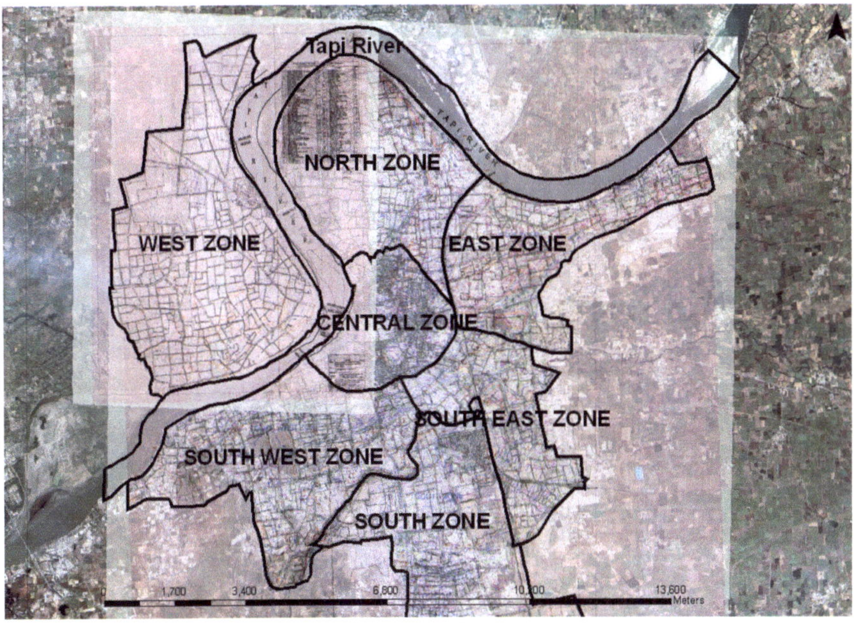

Fig. 8 Geo-referenced contour maps over Google earth image

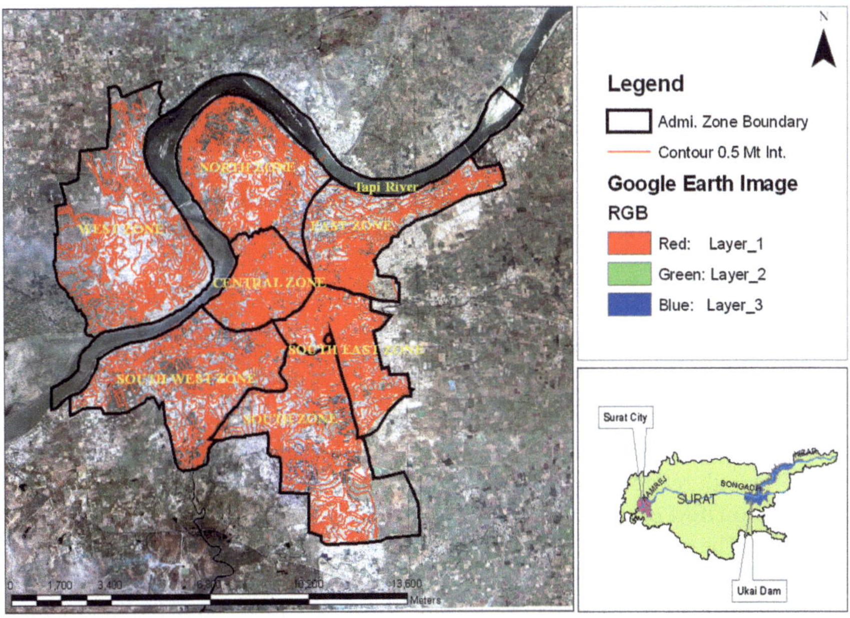

Fig. 9 Digitized contour overlying on Google earth image of Surat city

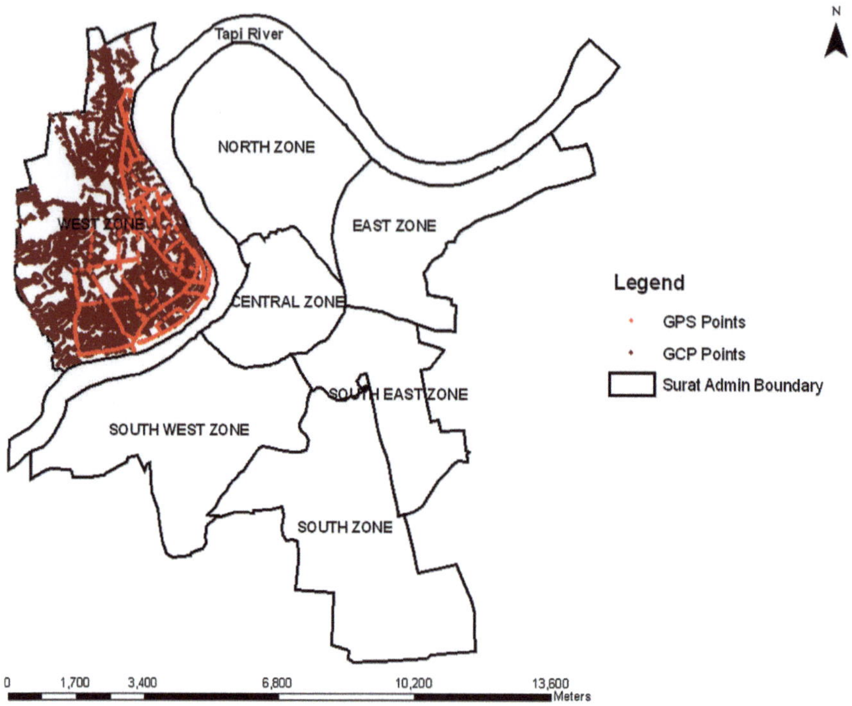

Fig. 10 GPS points and digitized contour (*GCP*) of west zone area

(Fig. 11). To find submerge areas in West Zone surface analysis is done using again 3D analyst tool. Arcscan is then applied for better visualization. For various flood potential values, areas are delineated and findings are analyzed from thematic flood inundation map for West Zone. Graph of submergence area of West zones according Town Planning Scheme (TPS) versus water level is then prepared.

4 Results and Discussion

As discussed earlier, DEM of the West Zone was generated by GPS points and collected contour map of 0.5 m interval. By considering the safe river gauge level of 7.55 m with reference to lowest bank height near Hope bridge and by applying this limit to DEM of the study area, flood inundation map is generated and shown in Fig. 12. The whole zone is segregated into 4 levels of submergence started with V1 (0–4 m RL) for very high inundation covering 0.16 km^2 area, V2 (4–8 m RL) for high inundation covering 15.15 km^2 area, V3 (8–12 m RL) for medium inundation covering 8.13 km^2 area and V4 (12 m and above) for low inundation covering 0.03 km^2 area. During the flood event 2006, observed river gauge level

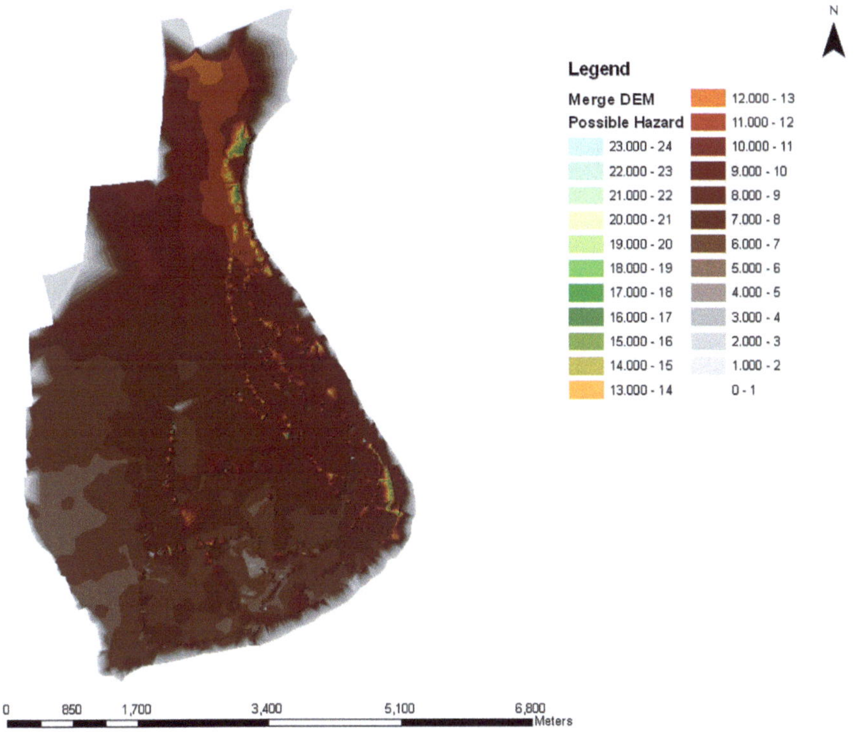

Fig. 11 Mearge TIN-DEM of west zone

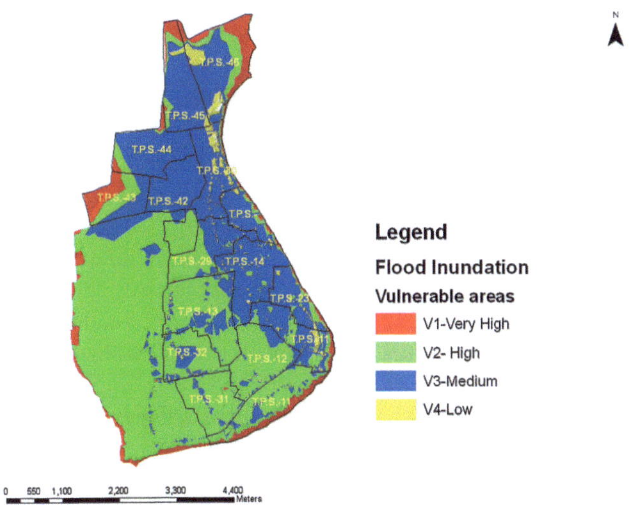

Fig. 12 Flood inundation and possible vulnerable areas of west zone Surat city

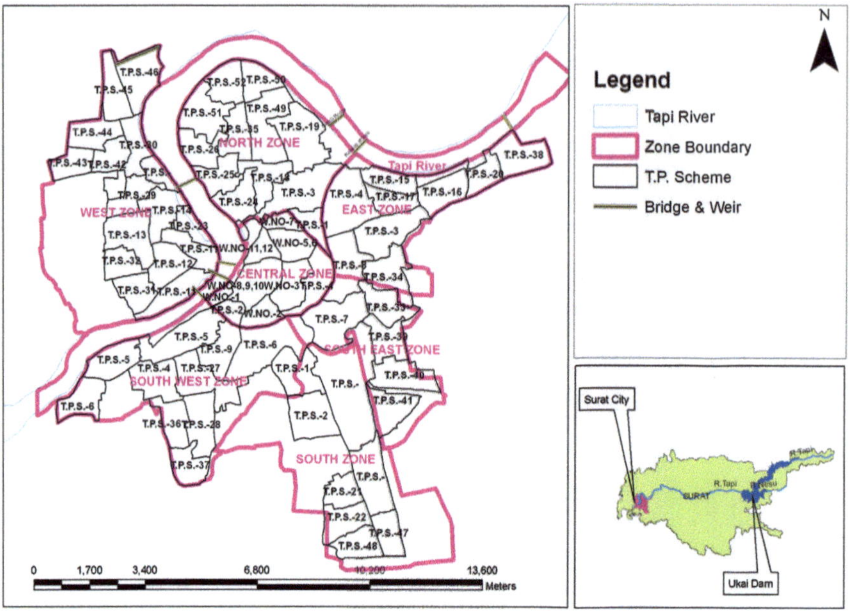

Fig. 13 Surat city map according to town planning scheme (*TPS*)

near Hope Bridge was 12.50 m. Flood inundation map indicates that areas of West Zone having reduced level above 12.50 m are safe and can survive under the same flood conditions. By comparison of the observed and calculated water levels of 2006 flood at Hope (Nehru) Bridge, it is found that the maximum water surface level in the west zone was at 4.91 m. This value is derived by subtracting lowest bank Reduced Level (RL), having value 7.55 m from 12.50 m river gauge level at Hope Bridge. For the forthcoming flood event, prediction of possible submergence can be done with reference to river gauge level, if the other conditions will remain same. By this comparison, graph of stage versus submerge areas for various TPS schemes of West Zone is plotted in which slopes of curves represent the flood inundation. In West Zone, foremost areas are Rander, Adajan, Jahagirabad-Pisad and Jahagirpura. These four renowned areas falls under different town planning scheme as intensify improvement of West Zone. Adajan was developed under TPS No. 11, 11S, 12, 13, 14, 31 and 32, Rander was developed under TPS No. 14, 23, 29 and 30 (Fig. 13, Table 2) while Jahangirabad is in TPS No. 42, 43, 44 and 45. Area of Adajan which falls under T.P. Scheme No.13 is having steep slope and it is a low lying area of this zone. Its 2.0 km^2 area was submerged when water level increased from 6 to 9 m. Similarly, area falling under TPS No. 11 S was affected at 8 m water level. Rander–Adajan area which belongs to TPS No. 14 is also a low level area and was affected by flood even at water level of 9–12 m and its 1.46 km^2 (Fig. 14) area was in submerge. Jahangirabad–Pisad, falls under T.P. Scheme No. 45, is developed at higher contour level and was affected when water

Table 2 Inundation area (m^2) of west zone as per the water level in m

Level in m	TPS no															
	Rander	T.P.S 11	T.P.S 11 S	T.P.S 12	T.P.S 13	T.P.S 14	T.P.S 23	T.P.S 29	T.P.S 30	T.P.S 31	T.P.S 32	T.P.S 42	T.P.S 43	T.P.S 44	T.P.S 45	T.P.S 45
	Rander	Adajan	Adajan	Adajan	Adajan	Rander-Adajan	Rander	Rander	Rander	Adajan	Adajan	Jahangirabad	Jahangirabad	Jahangirabad	Jahangirabad-Pisad	Jahangirpura
4	0	0	0	0	0	0	0	0	0	17,457	0	0	0	0	0	0
5	0	0	0	13,159	669	0	0	0	0	76,424	0	0	0	0	0	0
6	0	0	389,877	114,294	92,399	0	0	0	0	385,478	14,881	0	0	0	0	0
7	0	99,790	832,803	849,146	1,133,808	0	0	191,337	0	861,850	683,174	140,477	0	0	0	0
8	162,467	676,244	903,728	1,237,966	1,993,392	0	208,709	859,829	0	1,001,982	854,180	604,284	0	0	0	0
9	426,419	1,062,928		1,274,932	2,008,571	1,405	644,819	1,155,299	0			639,004	34,616	82,412	0	0
10	537,733	1,067,319				492,826	656,175		52,787			1,063,564	44,058	892,910	0	0
11	547,591					1,461,898			289,565			1,153,472		1,013,890	769,184	454,783
12						1,462,655			498,500						916,723	860,154
13									923,073						917,658	889,874
14									1,213,437							

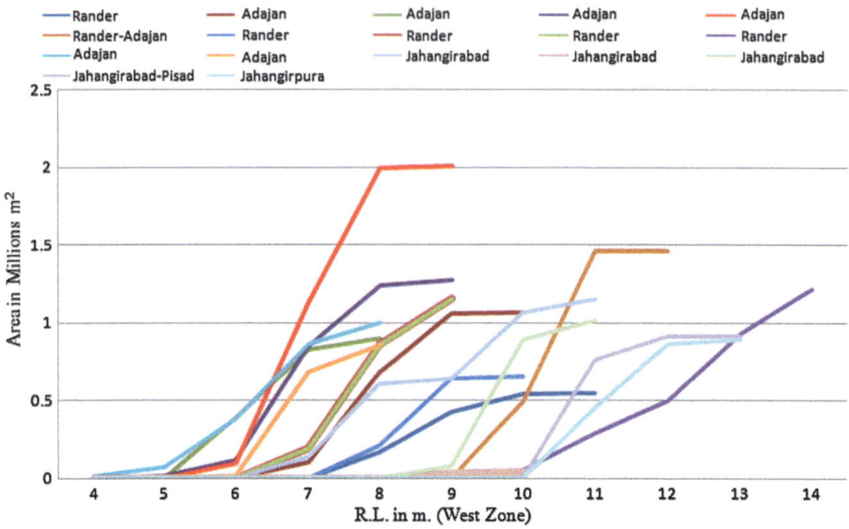

Fig. 14 Stage-area submerge curve of west zone and its TPS of Surat

level reached to 11 m. As per analysis, its about 0.92 km^2 area was in submerge. Jahngirpura follows the same characteristics as of Jahngirabad and is also a high rise area. At 13 m water level, its about 0.88 km^2 area was under water. According to the graph, Rader, Adajan are low laying areas in West Zone and for water level of 12.50 m, they were fully submerged. Comparison of this graph with the actual flood event 2006 shows that 90–95 % areas of the West Zone were submerged under water, which ultimately justify the obtained flood inundation map. It should be noted that the major parts of Adajan area, a part in the West Zone, was under water with water depth about 2–4 m.

5 Validation

For validation, Rander area is considered as case. Calculated data of inundation using software and actual observed field data are compared. As described in result, rander area was started to get affected from 4 m level and gone into submerge as level augmented. At LIC office, Rander branch, observed level was 3.65 m (12 ft.). DEM results shows that the reduced level of Rander was 8.5 m and observed gauge height at Nehru Bridge was 12.50 m and the difference between these heights is 4.0 m. For cross verification, the photographs of LIC office, Rander branch are used which indicates that the water height was about 12 ft. Noted reduced level of river section near Nehru bridge is 7.59 m. Thus, it is clear that the prediction for possible submergence of West Zone by DEM is nearby to actual observed field data.

6 Conclusion

GIS provides supplementary data in Hydrology for analysis while Digital Elevation Model (DEM) can be effectively used for simulation to get a complete model of the area. The analysis shows that West Zone is highly flood prone zone. Consequence of flood inundation was that 90–95 % area of West Zone, Surat, was under water. A steep slope in Stage-area submerge curve represents high potential of inundation while gentle slope represents a low potential. As West Zone is having low RL, it has more chances to get flooded severally. As per TPS, Rander and Adajan are low rise areas and will go under submerge at initial level of flooding. Certainly, application of Geo-spatial techniques such as RS, GIS and GPS would be the most efficient for flood inundation mapping.

Acknowledgments The authors would like to express their sincere thanks to Bhaskaracharya Institute For Space Applications and Geo-Informatics, National Bureau of Soil Survey and Land Use Planning, National Resources Information System, Survey of India, Central Water Commission, Irrigation Department, Surat Municipal Corporation and Technical Bulletin-Report on Reconnaissance Soil Survey of Surat District for providing necessary data, facilities and support during the study period.

References

Agnihotri P, Patel J (2008) Preparation of flood reduction plan for Surat city and surrounding region (India). Int J Trans Fluid Mech 2(3):116–125

Andrysiak PB, Maidment DR (2000) Visual floodplain modeling with geographical informatioain modeling with geographical information systems (GIS). The University Of Texas at Austin, Austin

Bates P, Marks K, Horritt M (2003) Optimal use of high-resolution topographic data in flood inundation models. Hydrol Process 17:537–557

Casale R, Margottini C (1999) Floods and landslides: integrated risk assessment. Springer, Berlin, pp 147–189

CWC (2000–2001) Water year book 2000–2001, Tapi Basin, Hydrological observation circle, Gandhinagar, Gujarat, India

Horritt M, Bates P (2002) Evaluation of 1D and 2D numerical models for predicting river flood inundation. J Hydrol 268:87–99

Horritt M, Bates P (2001) Predicting floodplain inundation: raster-based modelling versus the finite-element approach. Hydrol Process 15:825–842

Horritt M, Mason D, Luckman A (2001) Flood boundary delineation from synthetic aperture radar imagery using a statistical active contour model. Int J Remote Sens 22:2489–2507

Islam MDM, Sado K (2000) Development of flood hazard maps of Bangladesh using NOAA-AVHRR images with GIS. Hydrol Sci J 45:337–355

Joshi PM, Sherasia NK, Patel DP (2012) Urban flood mapping by geospatial technique a case study of Surat city. IOSR J Eng (IOSRJEN) 2:43–51

Kresch DL, Mastin M, Olsen T (2002) Fifty-year flood-inundation maps for Olanchito, Honduras, US Geological Survey, Tacoma, Washington, USA. The Survey

Neal J, Villanueva I, Wright N, Willis T, Fewtrell T, Bates P (2012) How much physical complexity is needed to model flood inundation? Hydrol Process 26:2264–2282

Noman NS, Nelson EJ, Zundel AK (2003) Improved process for floodplain delineation from digital terrain models. J Water Resour Planning Manage 129:427–436

Noman NS, Nelson EJ, Zundel AK (2001) Review of automated floodplain delineation from digital terrain models. J Water Resour Planning Manage 127:394–402

Patel DP, Dholakia MB (2010) Identifying probable submergence area of Surat city using digital elevation model and geographical information system. World Appl Sci J 9:461–466

Patel DP, Dholakia MB, Naresh N, Srivastava PK (2012a) Water harvesting structure positioning by using Geo-visualization concept and prioritization of mini-watersheds through morphometric analysis in the lower Tapi Basin. J Indian Soc Remote Sens 40:299–312

Patel DP, Gajjar CA, and Srivastava PK (2012b) Prioritization of Malesari mini-watersheds through morphometric analysis: a remote sensing and GIS perspective. Environ Earth Sci, pp 1–14

Patel DP, Srivastava PK (2013) Flood hazards mitigation analysis using remote sensing and GIS: correspondence with town planning scheme. Water Resour Manage, pp 1–16

Pistocchi A, Mazzoli P (ed) (2002) Use of HEC-RAS and HEC-HMS models with ArcView for hydrologic risk management. In: IEMS 2002 proceeding international environmental modelling and software society conference, Lugano, Switzerland, pp 305–310

Sadrolashrafi SS, Mohamed TA, Mahmud ARB, Kholghi MK, Samadi A (2008) Integrated modeling for flood hazard mapping using watershed modeling system. Am J Eng Appl Sci 1:149–156

Sulebak J (2000) Applications of digital elevation models. DYNAMAP Project, Oslo

Sutanta H (2002) Spatial modeling of the impact of land subsidence and sea level rise in a coastal urban setting, case study: Semarang, Central Java, Indonesia (Thesis). Type, M.Sc. thesis, International Institute for Geo-Information and Earth Observation, ITC, Enschede, The Netherlands

Tralli DM, Blom RG, Zlotnicki V, Donnellan A, Evans DL (2005) Satellite remote sensing of earthquake, volcano, flood, landslide and coastal inundation hazards. ISPRS J Photogrammetry Remote Sens 59:185–198

Wright N, Villanueva I, Bates P, Mason DC, Wilson M, Pender G et al (2008) Case study of the use of remotely sensed data for modeling flood inundation on the River Severn, UK. J Hydraulic Eng 134:533–540

Spatial Variations in Vegetation Fires and Carbon Monoxide Concentrations in South Asia

Krishna Prasad Vadrevu, Kristofer Lasko and Chris Justice

Abstract Vegetation fires are an important source of air pollution in several regions of the world including Asia. An important question with respect to satellite retrievals of air pollutants is "how well do they capture temporal and spatial variations and how well do they relate to episodic events such as fires?" We addressed this question using MOPITT surface CO and MODIS fire retrievals. We also evaluated MODIS aerosol optical depth (AOD) as well as small mode aerosol fraction (SMAF) variations in relation to fire seasonality. Results from temporal analysis (2003–2012) of fires in Asia suggested 22 % of all fires occurring in Myanmar, followed by India (20.91 %), Indonesia (18.31 %), Thailand (9.42 %), etc. Fire frequencies were highest in northeast India and Southeast Asia countries. Further, we observed significant spatial variation and seasonality in fires in Southeast Asia. In the northern Southeast Asia, the peak fire season was during January–March whereas in the south, the fires peak is from August through October. AOD followed a similar trend as that of fires, however, small mode aerosol fraction showed some discrepancies. Locally weighted regression yielded good results between vegetation fires and CO emissions. Results showed that areas with high vegetation fires were also areas of high CO emissions, with highest spatial correlation during the month of March. Among the fire counts and FRP, the correlations varied for individual months, however, both showed significant ($P < 0.001$) positive correlations suggesting that either of them can be used as predictor of CO concentrations. Locally weighted regression maps revealed how the relationship between fire counts versus CO and FRP versus CO change across time and space. The study captures the influence of vegetation fires on CO pollution in Asia using satellite data.

Keywords Vegetation fires · MODIS · MOPITT · Satellites · Carbon Monoxide · South Asia

K. P. Vadrevu (✉) · K. Lasko · C. Justice
Department of Geographical Sciences, University of Maryland, College Park, USA
e-mail: krisvkp@umd.edu

P. K. Srivastava et al. (eds.), *Remote Sensing Applications in Environmental Research*, Society of Earth Scientists Series, DOI: 10.1007/978-3-319-05906-8_8,

1 Introduction

Vegetation fires have significant regional and global impacts on the tropospheric chemistry and air pollution. In addition to slash and burn from forests (Prasad et al. 2008), agricultural residue burning to clear crop waste is another major source of air pollution in the Asian region (Li et al. 2010; Miettinen et al. 2011; Vadrevu et al. 2011, 2013). Historically, fires in the Asian region attracted international attention during 1997/1998 when a severe haze episode occurred due to wildfires in Indonesia. In the Indonesia, the area burned during 1997/1998 is estimated at 9.7 million hectares of forest and non-forest land, with some 75 million people affected by smoke, haze, and the fires themselves (Murdiyarso and Adiningsih 2007). The haze covered the entire area of Sumatra and Kalimantan and extended up to Malaysia, Singapore and the Southern part of Thailand due to long-range transport (Page et al. 2009; Bonnet and Garivait 2011). Impacts included damage to health, loss of life, property and reduced livelihood options. The economic costs were estimated to exceed 9 billion USD (Dennis et al. 2005). The impacts of the atmospheric pollution on health, transport and tourism, largely were borne by Indonesia, Brunei Darussalam, Singapore and Malaysia (Radojevic 2003). Fires in the Asian region were also associated with carbon emissions, environmental and economic losses (Vadrevu and Badarinath 2009; Vadrevu et al. 2010). For example, recent estimates suggest that the peat lands of southeast Asia cover approximately 250,000 km^2 (Page et al. 2011) and represent an immense reservoir of fossil carbon (Page et al. 2002; Jaenicke et al. 2008). CO_2 emissions from Peat fires are estimated at about $\sim$30 % of the global CO_2 emissions (Hooijer et al. 2012). Similarly, in India, the Himalayan forest fires during 1995 consumed nearly 6 Mha and caused huge economic losses (Kimothi and Jadhav 1998). The regular agricultural residue burning events that occur till date in the Punjab province of India were shown to cause enormous pollution problems including enhancing of direct radiative forcing in the Indo-Ganges region (Ramanathan and Carmichael 2008; Vadrevu et al. 2010). Likewise, Chan et al. (2000) from the ozonesonde profiles, fire count, Carbon monoxide (CO) data and back air trajectory analysis showed that biomass burning emissions in the south east Asia are the source of ozone enhancement in the lower troposphere in springtime at Hong Kong. More examples of the impacts of fires on the local environment specific to Asia can be found in Chan et al. (2000), Hsu et al. (2003), Ramanathan et al. (2005), Lau et al. (2010), Vadrevu et al. (2011), (2012a, b), Fu et al. (2012), etc. In addition to these climate impacts, adverse health impacts have been reported due to biomass burning pollutants. The aerosols released from biomass can cause asthma, bronchitis, emphysema, or pneumonia (Seaton et al. 1995; Nawahda et al. 2012). Carbon monoxide released from biomass burning can damage the respiratory system by interfering with the blood's ability to absorb oxygen (Suji et al. 1990). Considering the climate and health impacts of vegetation fires, accurate information on fires (location, frequency, extent, seasonality) is necessary to make informed decisions on the fire management and resulting ecological and air pollution impacts.

Links between biomass burning and air pollution need thorough investigation. In particular, CO is one of the dominant greenhouses released due to incomplete combustion of either fossil fuels or biomass. CO also has a relatively long atmospheric lifetime on the order of months and strongly influences the abundance of the OH radical, thereby altering the lifetime of methane and other greenhouse gases (Montzka et al. 2011). It is estimated that biomass burning accounts for about 331.1–351.52 Tg CO/yr with large inter-annual variations (van der Werf et al. 2006; Kaiser et al. 2012). Surface CO measurements over South Asia are extremely sparse and measurements of the vertical distribution of CO do not exist over the region (Kumar et al. 2013). Specific to the Asian region, biomass burning contributes to ~62–420 Tg C/yr (Streets et al. 2003; Hoelzemann et al. 2004; Van der Werf et al. 2006). Recently, Chang and Song (2010) using L3JRC and MCD45A1 satellite burned area products and following the Seiler and Crutzen approach (1980) estimated ~8.6 and 6.3 Tg CO/yr from forest/shrub and grassland burning compared to an earlier estimate of 35 Tg CO/yr (Streets et al. 2003). Chang and Song (2010) using satellite data also estimated that crop residue burning accounts for about 2.9–3.0 Tg/yr compared to 35 Tg CO/yr (Streets et al. 2003). These estimates clearly suggest significance of vegetation fires in releasing CO and also some discrepancies in the CO emission estimates.

The use of satellite remote sensing for characterizing the greenhouse gases in the troposphere is a science that has developed within the past twenty years (Burrows et al. 2011). For air quality mapping and monitoring, direct measurements of CO from space might resolve some uncertainties in emission estimates. CO has a strong absorption in the thermal infrared (4.7 μm) and shortwave infrared (2.3 μm) region of the spectrum. Based on this a number of space-borne instruments have been measuring tropospheric CO globally over the past decade, including MOPITT, SCIAMACHY, AIRS, ACE-FTS TES and IASI and as a result, the measurement of trace gas emissions from satellites has improved considerably (Kopacz et al. 2010). An important question with respect to satellite retrievals of air pollutants is "how well do they capture temporal and spatial variations and how well do they capture emissions from episodic events such as fires?" In this work, we address this question through analyzing the spatial variability in CO and vegetation fires in the Asia using MOPITT and MODIS datasets. Nine years of vegetation fire data has been used in conjunction with MOPITT data to characterize fire–CO relationships spatially and temporally. We focused on the below questions:

(a) What is the influence of vegetation fires on CO emissions in the Asian region? (b) How do active fires and fire radiative power (FRP) relate to CO emissions? (c) Which product, active fires or FRP better relates to CO emissions? (d) What are the spatial patterns in correlation strength between vegetation fires and CO emissions in the Asian region? (e) Does the correlation strength vary among different months, if so by how much? (f) How does aerosol optical depth and aerosol small mode fraction vary in relation to fires and how do they relate to CO?

In this study, in addition to fire counts, we also used FRP as a predictor of CO as FRP has been shown to directly relate to fire intensity and biomass consumption (Wooster et al. 2004; Ichoku et al. 2008). Thus, it was hypothesized that FRP should be well correlated with surface CO emissions retrieved from the MOPITT datasets.

2 Data Sets and Methodology

2.1 Vegetation Fires

For characterizing the vegetation fires in the region, we used daily active fire detections from the MODIS instruments onboard the Aqua and Terra satellites. The two MODIS sun-synchronous, polar-orbiting satellites pass over the Equator at approximately 10:30 a.m./p.m. (Terra) and 1:30 p.m./a.m. (Aqua) with a revisit time of 1–2 days. MODIS Advanced Processing System (MODAPS) processes the resulting data using the enhanced contextual fire detection algorithm (Giglio et al. 2003) combined into the Collection 5 Active Fire product. For this study, we analyzed the data from 2003–2012. The fire data are at 1 km spatial resolution at nadir; however, under ideal conditions it can detect flaming fires as small as 50 m^2. FRP is the rate of fire energy released per unit time, measured in megawatts (Kaufman et al. 1998). The MODIS algorithm for FRP is calculated as the relationship between the brightness temperature of fire and background pixels in the middle infrared (band center near 4 μm). It is given as (Kaufman et al. 1998),

$$FRP = 4.34 \times 10^{-19}(T_{MIR}^{8} - T_{bgMIR}^{8})$$

where FRP is the rate of radiative energy emitted per pixel, 4.34×10^{-19} MW km^{-2} K^8 is the constant derived from the simulations, T_{MIR} (Kelvin) is the radiative brightness temperature of the fire component, T_{bgMIR} (Kelvin) is the neighboring non-fire background component, and MIR refers to the middle infrared wavelength here, 3.96 μm. In this study, we utilized the Collection 5 Terra and Aqua monthly climate modeling grid datasets (MOD14CMH/MYD14CMH) that represent cloud and overpass corrected fire pixels data along with the mean FRP data.

2.2 MOPITT CO Retrievals

For characterizing the geographical extent of CO pollution, we used the MOPITT CO datasets. MOPITT instrument, launched in December 1999 aboard the NASA EOS Terra satellite, is a thermal nadir-viewing gas correlation radiometer (Drummond 1992; Deeter et al. 2003). It measures CO at a spatial resolution of

22 km × 22 km. Terra has a near polar sun-synchronous orbit with a descending equator crossing time of approximately 10:30 a.m. local time (ascending 10:30 p.m.). CO profiles are retrieved from radiance measurements in the thermal infrared channel at 4.7 mm (Deeter et al. 2003) and MOPITT achieves a global coverage within 3 days. We used the Version 4 Level 3 (L3) product which includes CO concentrations, averaging kernels and error co-variances at 10 pressure levels retrieved in linear space in surface, 900, 800, 700, 600, 500, 400, 300, 200, and 100 hPa) (Deeter et al. 2009). Of these columnar data, we used only surface data (~900 hPa) to infer fire–CO variations. Retrieved CO total columns are calculated by integrating the retrieved mixing ratio profile and each retrieval 'level' corresponds to a uniformly weighted layer immediately above that level. For example, the surface-level retrieval product corresponds to the mean volume-mixing ratio over the layer between the surface and 900 hPa. For CO vertical profiles, estimated errors are available in the error field of the "retrieved CO Mixing ratio profile" and "retrieved CO Surface mixing ratio" variables of the MOPITT files. These values represent the cumulative error from smoothing error, model parameter error, forward model error, geophysical noise, and instrument noise (Deeter and MOPITT team 2009) contributing to ~20–30 % uncertainty with 5 % difference in the monthly mean CO. A detailed review on the use satellite datasets for characterizing CO emissions from biomass burning has been reviewed by Monks and Beirle (2011).

2.3 Aerosol Optical Depth (AOD) and Aerosol Small Mode Fraction (SMAF)

We used the MODIS Collection 5.1 Terra (MOD08_M3) AOD at 550 nm (Remer et al. 2005; Levy et al. 2007) level 3 monthly product for characterizing the variations in relation to fires. The aerosol properties are derived by the inversion of the MODIS-observed reflectance using pre-computed radiative transfer look-up tables based on aerosol models (Remer et al. 2005; Levy et al. 2007). In addition, we also used the aerosol small/fine mode AOD fraction (SMAF) data at 500 nm for spatial variations in fire affected regions. SMAF is the ratio of small mode optical depth (thickness) to the total AOD.

2.4 Spatial Gridding, Ordinary Linear Regression (OLR) and Locally Weighted regression

We spatially gridded the vegetation fire data at 10-min intervals (~9.3 km^2 cells) to infer fire frequencies. We calculated the fire frequency per grid cell to see how many times an individual cell is impacted by fires over a period of nine years

(2003–2011). First, we summed the total number of fires reported by each data record $N_{k,m}^{MODIS,A,T}$ falling within a grid box, k, for each month (m) taking daily observations of Aqua and Terra MODIS fire counts, $M_{k,m}$ the number of months of MODIS observations (2003–2011). Then, we determined the frequency of fire per grid cell as $N_{k,m}^{MODIS,A,T} \big/ M_{k,m}$ for each month. The resulting map thus had a range of values from 0–9, with zero values suggesting that no fires occurred during the assessed period while a maximum of 9 suggesting that the pixel is impacted by fires every year starting (2003–2011). In addition to fire frequencies, for the same time period we also gridded the FRP. For CO, for spatial gridding, we used 0.5×0.5 degree resolution, as the MOPITT CO original data was in 22×22 km^2. Corresponding fire counts and FRP data were retrieved from MODIS climate modeling grid product (CMG) (Giglio et al. 2006). Fire counts and FRP data were then used in a regression framework for assessing the relationships using an ordinary linear regression as well as locally weighted regression.

An ordinary linear regression can be expressed as (Huang and Leung 2002),

$$Y_i = a_0 + \sum_{k=1}^{P} a_k x_{ik} + \theta_i \quad i = 1, \ldots n$$

where Y is the dependent variable and is represented as a linear combination of independent variables X_k, $k = 1, 2, P$; $a_0, a_1 \ldots, a_p$ are the parameters and θ_i are independent normally distributed error terms with zero mean and constant variance. In the OLR model, the parameters are assumed to be the same across the study region, which may not be true because different locations might have different parameters.

In this study, to better account for the spatial variations and relationships (spatial non-stationarity) in Fire/FRP versus CO, locally weighted regression also known as geographically weighted regression has been used (Cleveland and Devlin 1988; Fotheringham et al. 2002). The locally weighted regression accounts for the spatial heterogeneity in responses to variables by estimating separate regression for each sample observation including the location of interest and other spatially weighted observations. The weights represent the adjacency effects for neighboring locations within a specified distance (or bandwidth). Following the assumption that more proximate locations are more alike, the weights decay with distance following a bi-square decay function for an adaptive kernel. When regression points and observation points are the same, one regression is estimated for each observation, allowing parameter estimates to vary across the sample space. The locally weighted regression model is specified as

$$y_i = \beta_{i0} + \beta_{i1} x_{i1} + \beta_{i2} x_{i2} + \cdots + \beta_{ik} x_{ik} + \varepsilon_i; \varepsilon_i \sim N(0, \sigma^2), \quad i = 1, 2, \ldots n,$$

Table 1 Annual fire counts for different countries in South Asia obtained through averaging eleven years (2002–2012) of MODIS aqua and terra datasets

Country	Fire counts	% occurrence
Afghanistan	234	0.074
Bangladesh	3,114	0.980
Bhutan	223	0.070
Brunei	32	0.010
Cambodia	25,921	8.156
India	66,479	20.918
Indonesia	58,198	18.312
Laos	31,759	9.993
Malaysia	4,334	1.364
Maldives	0	0.000
Nepal	2,268	0.714
Pakistan	7,104	2.235
Philippines	4,475	1.408
Myanmar	67,503	21.240
Sri Lanka	1,289	0.406
Thailand	29,959	9.427
Vietnam	14,907	4.691
Singapore	6	0.002

Percent values are also shown

where, the 'I' subscripts on the parameters indicate that there is a separate set of (k + 1) parameters for each of the n-observations. The parameter estimates are given as,

$$\hat{\beta}_i = (X^i W_i X)^{-1} X' W_i Y; \quad i = 1, 2, \ldots n$$

where, W_i is the n × n weight matrix whose off-diagonal elements are zero land the diagonal elements are the weights of each observation relative to I, i.e., W_i = diagonal ($w_{i1}, w_{i2}, \ldots w_{in}$). The optimal bandwidth distance used in each observations regression is determined by the Akaike Information Criterion (AIC) test. The results obtained from this approach were reported in addition to OLR results.

3 Results and Discussion

Annual MODIS fire counts obtained through averaging the data from 2002–2012 for the south Asian countries are shown in Table 1 along with the percent contribution. Of the different countries, 22 % of all fires occurred in Myanmar, followed by India (20.91 %), Indonesia (18.31 %), Thailand 9.42 %), Cambodia (8.15 %), etc. Temporal statistics of fire counts for different countries are shown in Fig. 1a, b. Of the different years, 2004 had the highest number of fire counts in the Asian region with 406,627 fires, 2009 (367,534 fires), 2007 (358,433) and others. Spatial patterns in MODIS Aqua and Terra fire occurrences for the period of

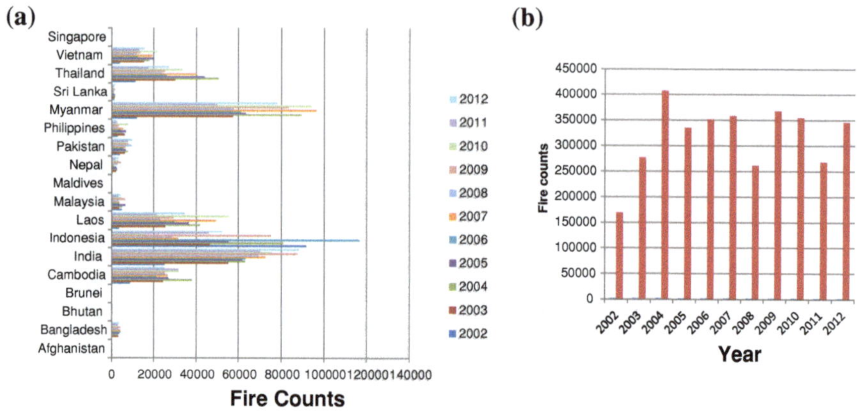

Fig. 1 **a** MODIS aqua and terra fire counts for individual countries (2002–2012). **b** MODIS total fire counts for different years

Fig. 2 **a** Fires over Punjab, India on October 22nd, 2012 and southeast Asia on May 11th, 2012. **b** MODIS aqua and terra fires were overlaid on MODIS aqua surface reflectance product (7:2:1)

October agricultural residue burning in Punjab and May forest burning for the study region are shown in Fig. 2a, b. The images clearly suggest significant amount of spatial variation in fires in Asia. The peak burning season is March with 3,702 fire counts per 10-min grid cells with a mean FRP (MW) of 226.28 MW (Figs. 3, 4). Although April recorded fewer fires (2,622) per 9.3 km^2 grid cell, it had a higher mean FRP (226.28 MW) than March (152.53 MW) suggesting relatively higher intensity (Fig. 4).

Fire frequency map calculated using nine years of MODIS data for the month of March at a 9.3 km^2 gridded data is shown in Fig. 5. Results clearly suggested highest fire frequencies in northeast India and southeast Asia countries, mainly Myanmar, Laos, Thailand, Vietnam, etc. Further, we observed significant spatial variation, seasonality as well as fire frequencies in Southeast Asia countries (Fig. 6a, b). For example, in the northern part of southeast Asia, the peak fire

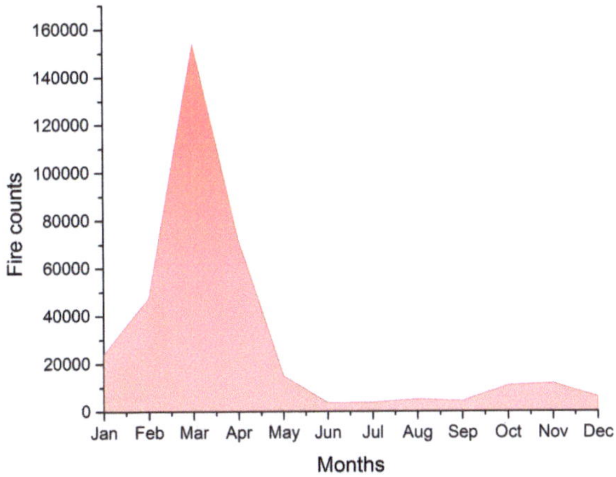

Fig. 3 MODIS aqua and terra monthly fire counts (averaged from 2002–2012)

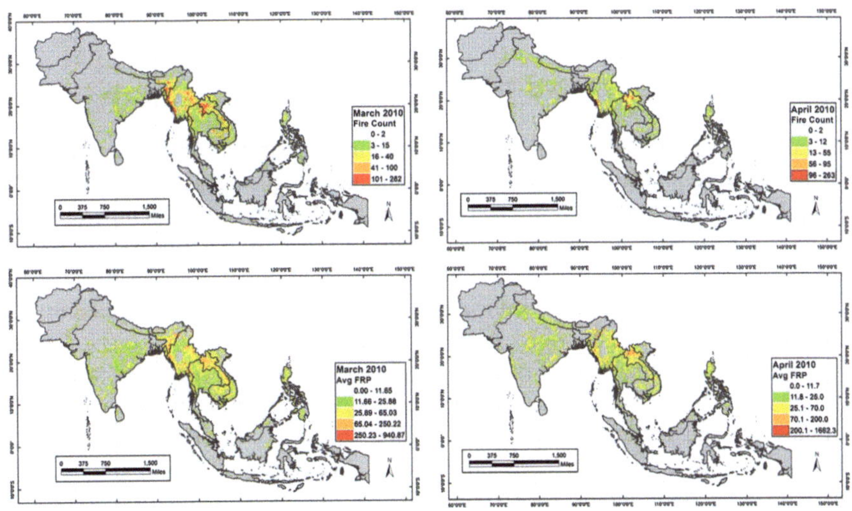

Fig. 4 MODIS fire counts and FRP (MW) aggregated at 9.3 km^2 grid cells (2010)

season was observed during January–March, especially in the region covering Cambodia, Laos PDR, Myanmar, Thailand and Vietnam; whereas in the southern southeast Asian countries of Indonesia and Malaysia, the peak burning season is from August through October (Fig. 6a, b). The fire seasonality has a significant impact on air pollution including CO concentrations.

CO data has been analyzed corresponding to fire counts in grid cells. CO concentrations peaked during March reaching above 700 ppbv during that month followed by April (625.6 ppbv). During winter, December had the highest CO concentrations (548.8 ppbv), followed by November (423.6 ppbv) (Table 2).

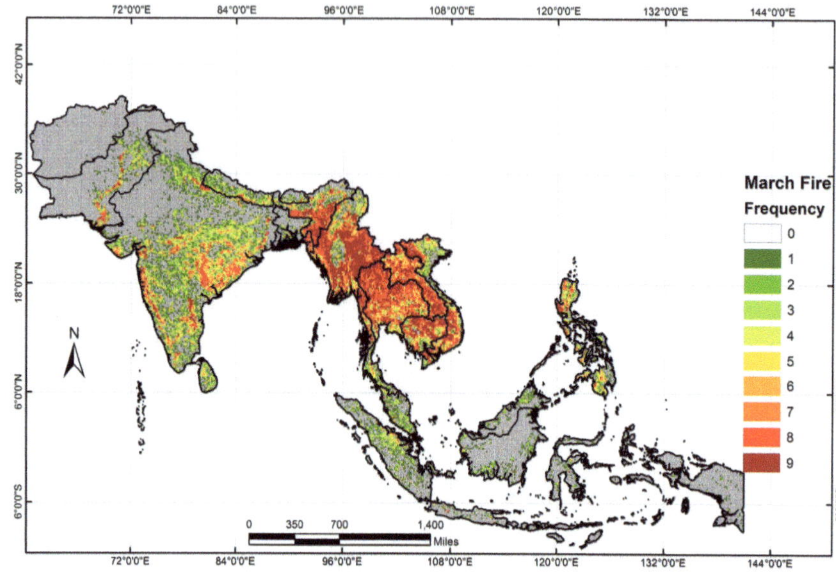

Fig. 5 Fire frequency per 9.3 km^2 grid cells during the month of March (2003–2011)

Spatial patterns suggested highest CO concentrations in the northeast India and Southeast Asia countries around 80–90° latitude and 20–30° longitude (Figs. 7, 8). In conjunction with the MODIS fire counts, AOD and SMAF showed temporal variations (Fig. 8). The seasonal trend in monthly average AOD suggested a steady increase of AOD from January (0.17) with a peak in March (0.61) and April (0.49) during summer and a second peak during October (0.22). Fine mode biomass burning aerosols were found to be quite high during Jan–March and also in November (Fig. 8). Although the Drastic increase in November SMAF coincides with agricultural residue burning (Gadde et al. 2011; Vadrevu et al. 2011), the increase could not be explained by fire counts and FRP alone (Fig. 8). One of the recent studies suggests that in addition to emission sources, relative humidity explains large increases or decreases in the SMAF (Bose 2012). Thus, more intense study is needed to evaluate SMAF variations in relation to meteorological parameters.

To assess the correlation strength between fire counts versus CO and also FRP versus CO, we tested the locally weighted regression approach in addition to OLR. We opted for locally weighted regression as the standard OLR approach estimates one fixed global set of regression coefficients. However, spatially clustered data such as fires and the resulting emissions (CO) could have residuals that can be either over or underestimated. In standard approaches, the ensuing spatial correlation caused by the underlying heterogeneity in the regression coefficients would be indistinguishable from standard spatial error correlation that is generated by clustered data (Fotheringham et al. 2002). Locally weighted regression can account for such errors and the results from fire counts versus CO and FRP versus CO for the peak biomass burning months were shown in (Fig. 9a, b and Table 3).

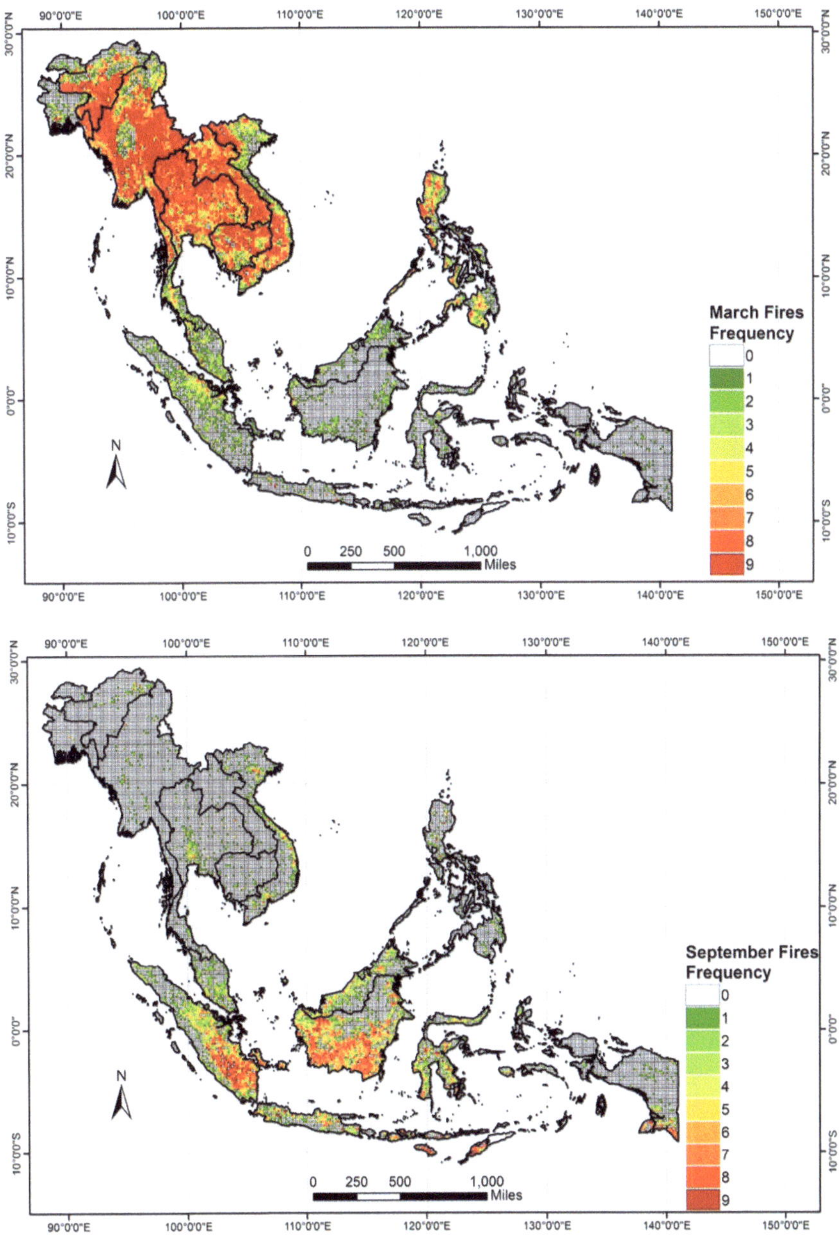

Fig. 6 **a** Spatial variation in fire frequencies in southeast Asia for the month of March and September (2003–2011). **b** A clear north–south gradient in fire frequencies and seasonality was observed in the region

Table 2 Monthly MOPITT surface CO (ppbv) statistics for 0.5 × 0.5 km^2 grid cells obtained from averaging 2002–2010 datasets over the South Asia region

Months	Minimum	Maximum	Mean	St. Dev
Jan	0	567.04	164.38	117.25
Feb	0	570.07	182.52	118.67
Mar	80.34	704.58	271.7	129.07
Apr	0	625.6	159	117.1
May	0	388.83	120.27	77.06
Oct	0	529.21	118.3	85.25
Nov	0	423.65	124.08	90.72
Dec	0	548.81	148.15	120.83

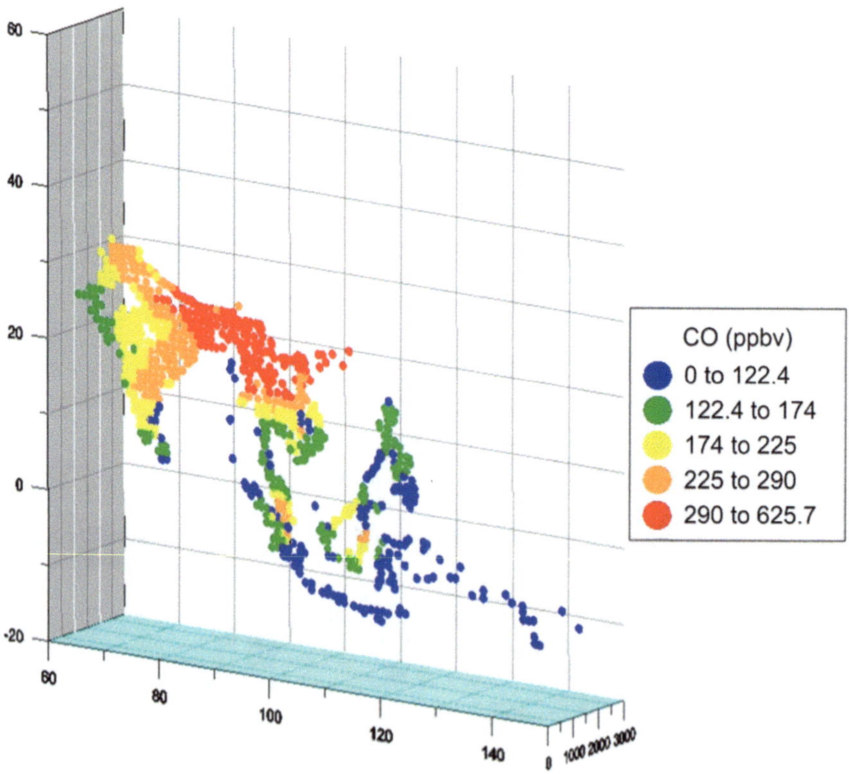

Fig. 7 MOPITT CO (ppbv) concentrations during the peak biomass burning month of March (2010)

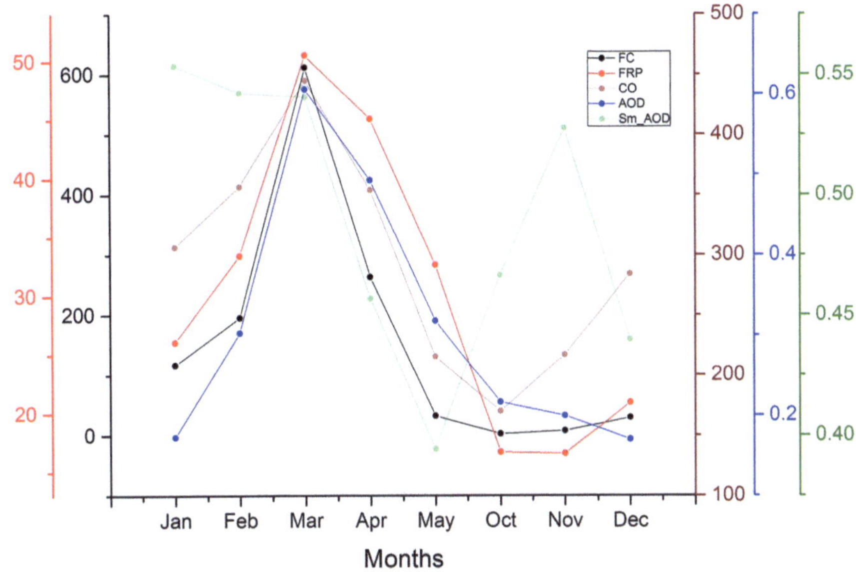

Fig. 8 Variations in fire counts, FRP (MW), CO (ppbv), AOD and small mode AOD in southeast Asia during the peak biomass burning months (2010). The values are mean per 9.3 km^2 grid cells averaged across southeast Asia

We observed a clear increase in "goodness of fit" and adjusted R^2 using the locally weighted regression approach compared to the OLR. For example, during the peak biomass burning month of March, fire counts versus CO using OLR showed r^2 of 0.37 while locally weighted regression showed r^2 of 0.82, almost 45 % increase in correlation and explanation of variance. Among the fire counts and FRP, the correlation strength varied for individual months, however, the relationship was strong and highly significant with both variables ($P < 0.001$) suggesting that either fire counts or FRP can be used as predictor of CO concentrations (Table 3). For example, for the peak biomass burning month of March, fire counts could explain 81 % of variance, whereas FRP could explain 77 %. For other months of January, February and April, both the fire counts and FRP had almost similar adjusted r^2.

Spatial correlation maps of Fire–CO and FRP–CO relationships suggested highest correlation over northeast India and Southeast Asia countries of Myanmar, Thailand, Laos, Vietnam, etc. The lowest correlation was evident over Rajasthan desert regions which didn't have fires and also over Pakistan, Afghanistan, west coast of India with very few fire counts, thus low CO concentrations. The fire counts versus CO and FRP versus CO spatial correlations were consistent with the location and intensity of MODIS fires. In general, when solving for emissions within a selected region or a grid cell, contribution from outside such as through long range transport needs addressing such as through 3D atmospheric transport modeling (Pfister et al. 2005; Stohl 2002). Even without transport modeling, using

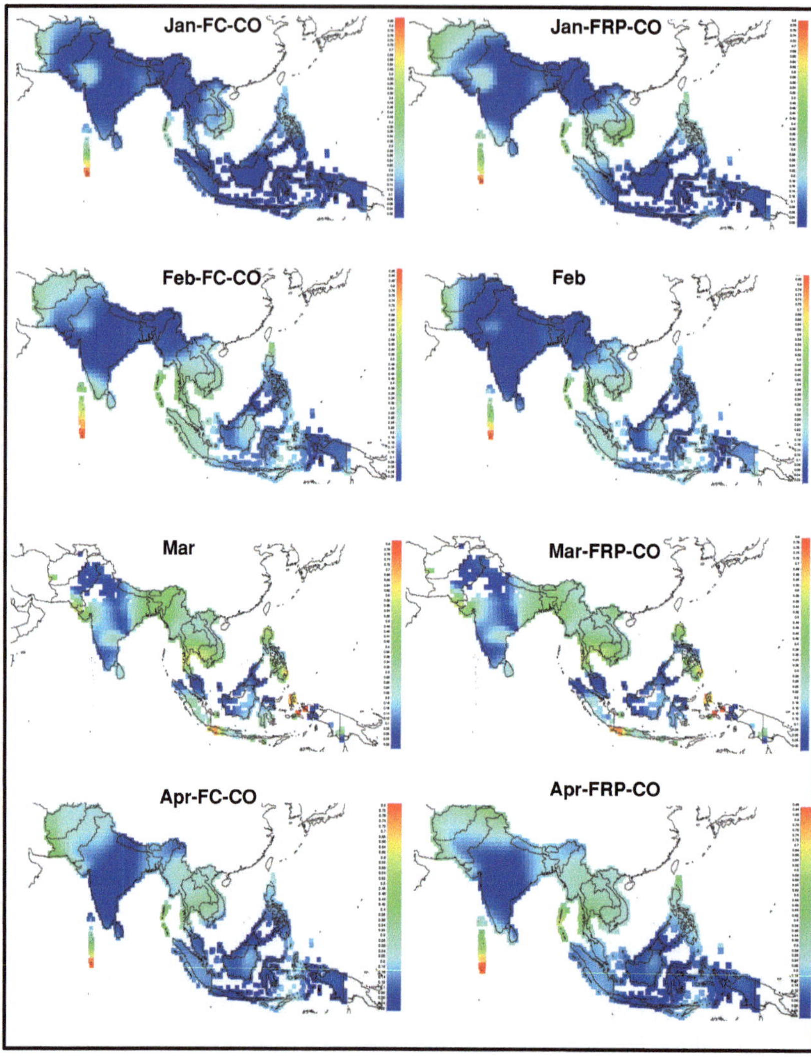

Fig. 9 **a** Spatial correlations between MODIS fire counts (FC), FRP and CO for individual months at 0.5 × 0.5 degree grid cells. **b** The correlation values from ordinary least square regression (OLR) and local regression are shown in Table 3

the locally weighted regression approach, we could explain 80 % of variance in CO datasets from fires during the peak biomass burning months. Our analysis clearly suggests biomass burning influence on CO concentrations, both spatially and temporally in the Asian region. Further, we used surface CO retrievals from MOPITT that are more representative of surface processes and emission sources in

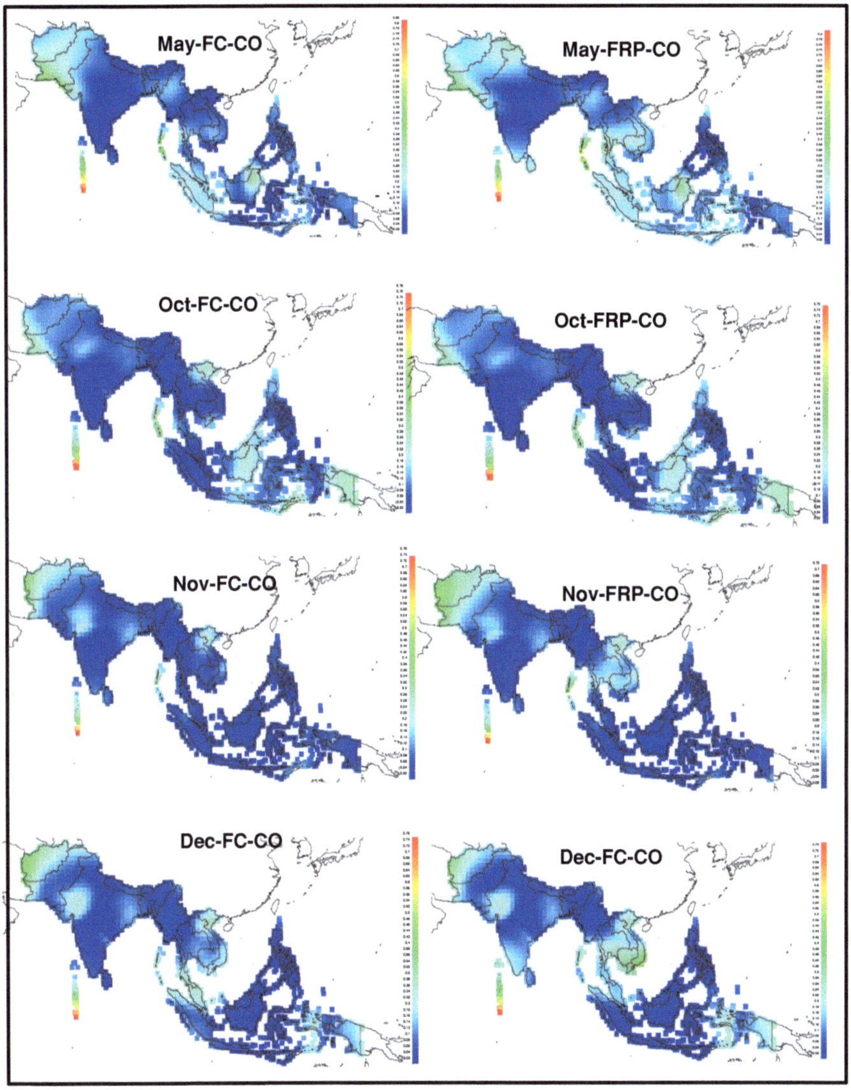

Fig. 9 continued

contrast to the total columnar measurements. Thus, our results also highlight the potential of surface level CO retrievals from MOPITT in capturing spatial and temporal variations and fire episodic events. In contrast, more work is needed to resolve discrepancies in SMAF variations and its' relationship with biomass burning and other emission sources.

Table 3 Regression results from (a) fire counts as predictor and CO as response variable for South Asia (b) FRP as predictor and CO as response variable for South Asia

Months	Correlation coefficient (r)	Coefficient of determination (r^2)	Adjusted r-square	*P*-value
(a)				
Jan	0.838 (0.167)	0.702 (0.028)	0.697 (0.028)	<0.001
Feb	0.834 (0.367)	0.696 (0.135)	0.69 (0.135)	<0.001
Mar	0.90 (0.60)	0.822 (0.37)	0.817 (0.37	<0.001
Apr	0.834 (0.429)	0.696 (0.184)	0.691 (0.184)	<0.001
May	0.758 (0.209)	0.575 (0.044)	0.568 (0.044)	<0.001
Oct	0.623 (0.09)	0.388 (0.008)	0.377 (0.008)	<0.001
Nov	0.8 (0.09)	0.64 (0.008)	0.634 (0.008)	<0.001
Dec	0.874 (0.182)	0.764 (0.033)	0.759 (0.033)	<0.001
(b)				
Jan	0.846 (0.335)	0.715 (0.112)	0.709 (0.112)	<0.001
Feb	0.84 (0.422)	0.705 (0.178)	0.699 (0.178)	<0.001
Mar	0.886 (0.374)	0.786 (0.14)	0.777 (0.14)	<0.001
Apr	0.842 (0.562)	0.709 (0.315)	0.703 (0.315)	<0.001
May	0.772 (0.386)	0.595 (0.149)	0.587 (0.149)	<0.001
Oct	0.633 (0.184)	0.401 (0.034)	0.389 (0.034)	<0.001
Nov	0.801 (0.061)	0.641 (0.004)	0.634 (0.004)	<0.001
Dec	0.876 (0.232)	0.768 (0.054)	0.763 (0.054)	<0.001

The standard OLS values are shown in parenthesis along with the local regression values. Results for June, July and August are not presented as most of the region is impacted by cloud cover

Acknowledgements The authors would like to thank Dr. Louis Giglio (UMd) for the MODIS active fire data set and CMG product used in this study. This research was supported by NASA grant NNX10AU77G.

References

Bonnett S, Gariavait S (2011) Seasonal variability of biomass open burning activities in the greater mekong sub-region. In: Tsuruta H, Fukuyama K (eds) Biomass burning and impacts impacts on Earth's environment vol 15(1). Global Environmental Research, pp 31–37

Bose S (2012) Increases and decreases in the fine mode fraction of aerosol optical depth with increasing relative humidity. Univ Illinois Urbana-Champaign: Thesis. http://hdl.handle.net/2142/29662

Burrows JP, Platt U, Borrell P (2011) The remote sensing of tropospheric composition from space. Springer, New York

Chan LY, Chan CY, Liu HY, Christopher S, Oltmans SJ, Harris JM (2000) Case study on the biomass burning in Southeast Asia and enhancement of tropospheric ozone over Hong Kong. Geophys Res Lett 27(10):1479–1482

Chang D, Song Y (2010) Estimates of biomass burning emissions in tropical Asia based on satellite-derived data. Atmos Chem Phys 10:2335–2351

Cleveland WS, Devlin SJ (1988) Locally weighted regression: an approach to regression analysis by local fitting. J Am Stat Assoc 83(403):596–610

Deeter MN, MOPITT Algorithm Development Team (2009) MOPITT (measurements of pollution in the troposphere) validated version 4 product user's guide. http://www.acd.ucar.edu/mopitt/v4_users_guide_val.pdf
Deeter MN, Emmons LK, Francis GL, Edwards DP, Gille JC, Warner JX, Khattatov B, Ziskin D, Lamarque JF, Ho SP, Yudin V, Attie JL, Packman D, Chen J, Mao D, Drummond JR (2003) Operational carbon monoxide retrieval algorithm and selected results for the MOPITT instrument. J Geophys Res 108(D14):4399. doi:10.1029/2002JD003186
Dennis RA, Mayer J, Applegate G (2005) Fire, people and pixels: linking social science and remote sensing to understand underlying causes and impacts of fires in Indonesia. Hum Ecol 33:465–504
Drummond JR (1992) Measurements of pollution in the troposphere (MOPITT). In: Gille JC, Visconti G (eds) The use of EOS for studies of atmospheric physics. New York, North-Holland, pp 77–101
Fotheringham A, Brunsdon SC, Charlton M (2002) Geographically weighted regression. Wiley, New York
Fu JS, Hsu NC, Gao Y, Huang K, Li C, Lin NH, Tsay SC (2012) Evaluating the influences of biomass burning during 2006 BASE-ASIA: a regional chemical transport modeling. Atmos Chem Phys 12:3837–3855
Gadde B, Bonnet S, Menke C, Garivait S (2011) Air pollutant emissions from rice straw open field burning in India, Thailand and the Philippines. Environ Pollut 157(5):1554–1558
Giglio L, Descloitres J, Justice CO, Kaufman Y (2003) An enhanced contextual fire detection algorithm for MODIS. Remote Sens Environ 87:273–282
Giglio L, Csiszar I, Justice CO (2006) Global distribution and seasonality of active fires as observed with the terra and aqua moderate resolution Imaging spectroradiometer (MODIS) sensors. J Geophys Res 111(G2):G02016
Hoelzemann JJ, Schultz MG, Brasseur GP, Granier C, Simon M (2004) Global wildland fire emission model (GWEM): evaluating the use of global area burnt satellite data. J Geophys Research 109(D14):D14S04
Hooijer A, Page S, Jauhiainen J, Lee WA, Lu XX, Idris A, Anshari G (2012) Subsidence and carbon loss in drained tropical peatlands. Biogeosciences 9(3):1053–1071
Hsu NC, Herman JR, Tsay SC (2003) Radiative impacts from biomass burning in the presence of clouds during boreal spring in Southeast Asia. Geophys Res Lett 30(5):1224
Huang Y, Leung Y (2002) Analyzing regional industrialization in Jiangsu province using geographically weighted regression. J Geogr Syst 4(2):233–249
Ichoku C, Giglio L, Wooster MJ, Remer LA (2008) Global characterization of biomass-burning patterns using satellite measurements of fire radiative energy. Remote Sens Environ 112(6):2950–2962
Jaenicke J, Rieley JO, Mott C, Kimman P, Siegert F (2008) Determination of the amount of carbon stored in Indonesian peatlands. Geoderma 147:151–158. doi:10.1016/j.geoderma.2008.08.008
Kaiser JW, Heil A, Andreae MO, Benedetti A, Chubarova N, Jones L, Morcrette JJ, Razinger M, Schultz MG, Suttie M, van der Werf GR (2012) Biomass burning emissions estimated with a global fire assimilation system based on observed fire radiative power. Biogeosciences 9(1):527–554
Kaufman YJ, Justice CO, Flynn L, Kendall JD, Prins EM, Giglio L, Ward D, Menzel W, Setzer A (1998) Potential global fire monitoring from EOS-MODIS. J Geophys Res 103(D24):32215–32238
Kimothi MM, Jadhav RN (1998) Forest fire in the Central Himalaya: an extent direction and spread using IRS LISS-I data. Int J Remote Sens 19(12):2261–2274
Kopacz M, Jacob DJ, Fisher JA, Logan JA, Zhang L, Megretskaia IA, Yantosca RM, Singh K, Henze DK, Burrows JP, Buchwitz M, Khlystova I, McMillan WW, Gille JC, Edwards DP, Eldering A, Thouret V, Nedelec P (2010) Global estimates of CO sources with high resolution by adjoint inversion of multiple satellite datasets (MOPITT, AIRS, SCIAMACHY, TES). Atmos Chem Phys 10(3):855–876

Lau WKM, Kim MK, Kim KM, Lee WS (2010) Enhanced surface warming and accelerated snowmelt in the Himalayas and Tibetan Plateau induced by absorbing aerosols. Environ. Res. Lett 5(2):025204. doi:10.1088/1748-9326/5/2/025204

Levy RC, Remer LA, Mattoo S, Vermote EF, Kaufman YJ (2007) Second-generation operational algorithm: retrieval of aerosol properties over land from inversion of moderate resolution imaging spectroradiometer spectral reflectance. J Geophys Res 112(D13):D13211

Li H, Han Z, Cheng T, Du H, Kong L, Chen J, Zhang R, Wang W (2010) Agricultural fire impacts on the air quality of Shanghai during summer harvest time. Aerosol Air Quality Res 10(2):95–101

Miettinen J, Shi C, Liew SC (2011) Deforestation rates in insular Southeast Asia between 2000 and 2010. Glob Change Biol 17(7):2261–2270

Monks PS, Beirle S (2011) Applications of satellite observations of tropospheric composition. In: Burrows JP, Platt U, Borrell P (eds) The remote sensing of tropospheric composition from space. Springer, Berlin, pp 365–449

Montzka SA, Krol M, Dlugokencky E, Hall B, Jöckel P, Lelieveld J (2011) Small interannual variability of global atmospheric hydroxyl. Science 331(6013):67–69

Murdiyarso D, Adiningsih ES (2007) Climate anomalies, Indonesian vegetation fires and terrestrial carbon emissions. Mitig Adapt Strat Glob Change 12(1):101–112

Nawahda A, Yamashita K, Ohara T, Kurokawa J, Yamaji K (2012) Evaluation of premature mortality caused by exposure to PM 2.5 and Ozone in East Asia: 2000, 2005, 2020. Water Air Soil Pollut 223:1–15

Page SE, Sigert F, Rieley JO, Boehm HDV, Jaya A, Limin S (2002) The amount of carbon released from peat and forest fires in Indonesia during 1997. Nature 420(6911):61–65

Page S, Hoscilo A, Langner A, Tansey K, Siegert F, Limin S, Rieley J (2009) Tropical peatland fires in Southeast Asia. Trop Fire Ecol, 263–287

Page SE, Rieley JO, Banks CJ (2011) Global and regional importance of the tropical peatland carbon pool. Glob Change Biol 17(2):798–818

Pfister G, Hess PG, Emmons LK, Lamarque J-F, Wiedinmyer C, Edwards DP, Pétron G, Gille JC, Sachse GW (2005) Quantifying CO emissions from the 2004 Alaskan wildfires using MOPITT CO data. Geophys Res Lett 32:L11809. doi:10.1029/2005GL022995

Pope CA III, Dockery DW (2006) Health effects of fine particulate air pollution: lines that connect. J Air Waste Manag Assoc 56(6):709–742

Prasad KV, Badarinath KVS, Anuradha E (2008) Biophysical and anthropogenic controls of forest fires in the Deccan Plateau, India. J Environ Manag 86:1–13

Radojevic M (2003) Chemistry of forest fires and regional haze with emphasis on Southeast Asia. Pure appl Geophys 160(1):157–187

Ramanathan V, Chung C, Kim D, Bettge T, Buja L, Kiehl JT, Washington WM, Fu Q, Sikka DR, Wild M (2005) Atmospheric brown clouds: impacts on South Asian climate and hydrological cycle. Proc Natl Acad Sci USA 102(15):5326-5333

Ramanathan V, Carmichael G (2008) Global and regional climate changes due to black carbon. Nat Geosci 1(4):221–227

Remer L et al (2005) The MODIS aerosol algorithm, products, and validation. J Atmos Sci 62(4):947–973

Seaton A, Godden D, MacNee W, Donaldson K (1995) Particulate air pollution and acute health effects. The Lancet 345(8943):176–178

Stohl A, Eckhardt S, Forster C, James P, Spichtinger N (2002) On the pathways and timescales of intercontinental air pollution transport. J Geophys Res 107(D23):4684. doi:10.1029/2001JD001396

Streets DG, Yarber KF, Woo J-H, Carmichael GR (2003) Biomass burning in Asia: annual and seasonal estimates and atmospheric emissions. Global Biogeochem Cycles 17(4):1099. doi:10.1029/2003GB002040

Sugi K, Theissen JL, Traber LD, Herndon DN, Traber DL (1990) Impact of carbon monoxide on cardiopulmonary dysfunction after smoke inhalation injury. Circ Res 66(1):69–75

Vadrevu KP, Badarinath KVS (2009) Spatial pattern analysis of fire events in central India—a case study. Geocarto Int 24(2):115–131

Vadrevu KP, Anuradha E, Badarinath. KVS (2010) Fire risk evaluation using multi-criteria analysis—A case study. Environ Monitoring Assess 166(1–4):223–239. doi:10.1007/s10661-009-0997-3

Vadrevu KP, Ellicott E, Badarinath KVS, Vermote E (2011) MODIS derived fire characteristics and aerosol optical depth variations during the agricultural residue burning season, North India. Environ Pollut 159(6):1560–1569

Vadrevu KP, Ellicott E, Giglio L, Badarinath KVS, Vermote E, Justice C, Lau KW (2012a) Vegetation fires in the Himalayan region—aerosol load, black carbon emissions and smoke plume heights. Atmos Environ 47:241–251

Vadrevu KP, Csiszar I, Ellicott E, Giglio L, Badarinath KVS, Vermote E, Justice C (2012b) Hotspot Analysis of Vegetation Fires and Intensity in the Indian Region. IEEE J Sel Top Appl Earth Obs Remote Sens 99:1–15

Vadrevu KP, Giglio L, Justice C (2013) Satellite-based analysis of fire–CO relationships from forest and agriculture residue burning (2003–2011). Atmos Environ. doi:10.1016/j.atmosenv.2012.09.055

Van der Werf GR, Randerson JT, Giglio L, Collatz GJ, Kasibhatla PS, Arellano AF Jr (2006) Interannual variability in global biomass burning emissions from 1997–2004. Atmos Chem Phys 6:3423–3441. doi:10.5194/acp-6-3423

Wooster MJ, Zhang YH (2004) Boreal forest fires burn less intensely in Russia than in North America. Geophys Res Lett 31(L20505):10–1029

Wooster MJ, Roberts G, Perry GLW, Kaufman YJ (2005) Retrieval of biomass combustion rates and totals from fire radiative power observations: FRP derivation and calibration relationships between biomass consumption and fire radiative energy release. J Geophys Res 110:D24311–D24311

Zhang L, Li QB, Jin J, Liu H, Livesey N, Jiang JH, Mao JH, Chen D, Luo M, Chen Y (2011) Impacts of 2006 Indonesian fires and dynamics on tropical upper tropospheric carbon monoxide and ozone. Atmos Chem Phys 11:10929–10946

Land Use Fragmentation Analysis Using Remote Sensing and Fragstats

Sudhir Kumar Singh, Avinash Chandra Pandey and Dharamveer Singh

Abstract This study present the results of a set of landscape metrics derived from remotely sensed data aiming to characterize the historical trends of landscape changes in the Allahabad district in the period 1990–2010. However, the identified trends in landscape changes and its effects in the region have potential policy implications. The land use and land cover were estimated from sensors viz. for the period 1990 (LANDSAT TM), 2000 (LANDSAT ETM^{+}) and 2010 (IRS 1D LISS III) through the maximum likelihood classification (MLC) method. The land use land cover change was quantified with the help of ERDAS imagine 9.1. Further, landscape level and class level metrics were derived from the classified satellite images in FRAGSTATS 3.3. Total four metrics for landscape level viz. total area (TA), number of patch (NP), patch density (PD), area mean (AREA MN) and four metrics for class level viz. core area (CA), number of patch (NP), patch density (PD) and percentage of land (PLAND), respectively to uncover the influence of land use change which can be correlated to the degree of urbanization, development and water quality. The different class level metrics of study area has revealed internal exchange of four land use classes given as agricultural land (65.32 % in 1990, 67.13 % in 2000, 68.1 % in 2010), builtup area (9.98 % in 1990,11.63 % in 2000,13.36 % in 2010), cultivable land (4.42 % in 1990, 3.47 % in 2000, 2.1 % in 2010) forest (6.03 % in 1990, 4.47 % in 2000, 5.6 % in 2010), and water body (5.89 % in 1990, 5.82 % in 2000, 5.35 % in 2010). The study showed that the notable changes had occurred in the last 20 years in this landscape, hence there is need of appropriate measures to mitigate these negative impacts of changes.

Keywords Landscape pattern · Fragmentation · LISS III · LANDSAT · Ecological metrics

---

S. K. Singh (✉) · A. C. Pandey · D. Singh
K. Banerjee Centre of Atmospheric and Ocean Studies, IIDS, Nehru Science Centre, University of Allahabad, Allahabad 211002, India
e-mail: sudhirinjnu@gmail.com

P. K. Srivastava et al. (eds.), *Remote Sensing Applications in Environmental Research*, Society of Earth Scientists Series, DOI: 10.1007/978-3-319-05906-8_9,

1 Introduction

Growing concerns over the recent urban and rural developmental activities, changing landscape scenario and the loss of biodiversity has made it an imperative for land managers to seek for better ways of managing the landscape and to be able to analyze positive and negative factors that influence the land use and land cover change pattern and its change dynamics. Land use land cover change is a continuous process which is altered by several ways with respect to time and space. The natural and socio–economic factors and their utilization by man in time have greatly affected historical land use and land cover pattern. Therefore, the accurate information pertinent to land use land cover is essential for the selection, planning and implementation of land use programmes which can be used to meet the increasing demands for basic human needs and their welfare (Zubair 2006). Rapid population increment and industrial sprawl has made land a scarce resource, mainly in urban and urbanizing areas which are undergoing to large–scale land transformation changes which alters the natural ecosystem in different way at temporal and spatial scale (Morley and Karr 2002). Change in land use land cover also aggravates the soil erosion, creates strong environmental impacts, effects on agricultural production, infrastructure and water quality (Lal 1998; Pimentel et al. 1995). The productive agricultural land and essential forest, neither of which can resist nor deflect the overwhelming land use and land cover but accelerate its change and alteration on the basis of necessity and sustainable development. This growth is an indicator of socio-economic development and generally has a negative enviornmental impact on the region. Agricultural practices, forest cutting, urban and industrial expansion, road development, military training, alteration of water way and other human activities have significant impact on land cover (Dale et al. 1998). Today one of the greatest challenge in front of human lies in the question of 'how to minimize the negative impacts of land use land cover change by applying both appropriate and cost-effective technology respectively?'. The landscapes metrics is a special feature that provide the ability of quantifying land use and land cover pattern distribution. There are variety of landscape metrics to allow quantitative assessment of a landscape and its level of fragmentation (McGarigal and Marks 1995).

The analysis of landscape level and class level metrics (Szabó et al. 2012) has provided a strong conceptual and theoretical basis for understanding landscape structure, function and change. The landscape metrics mainly focus on the three characteristics of landscape (Forman and Godron 1986; Turner 1989). (1) Structure: the spatial relationships among the distinctive ecosystems or elements present—more specifically, the distribution of energy, materials, and species in relation to the sizes, shapes, numbers, kinds, and configurations of the ecosystems. (2) Function: the interactions among the spatial elements, that is, the flow of energy, materials, and species among the component of ecosystems. (3) Change: the alteration in the structure and function of the 'Ecological Mosaic' over time. The landscape level metrics is a special feature of land use that provide significance ability to quantify the landscape patterns and the interactions among patch density, number of patches, total area, and large patch index within a landscape mosaic and also represent the patterns and interaction change over time. The class level metrics comprising with different metrics viz. core area, number of patches, patch density, percentage of landscape and largest patch index, some of these metrics quantify the landscape composition and while others quantify landscape configuration. It is also known that composition and configuration can affect ecological processes independently and interactively. So it is necessary to understand each metrics of landscape.

2 Study Area

This study area Allahabad district is situated at (24°47′N and 25°47′N to 81°19′E and 82°21′E) and it covers an area 5,246 km^2. It lies in the northern-central part of the densely populated eastern region of state Uttar Pradesh, in between the Gangetic plain and adjoining Vindhyan plateau of India. River Ganga and Yamuna flowing through the district (Fig. 1 location map of the study area). Geologically the district presents a greater complexity, the whole Trans-Ganga tracts, the greater portion of 'doab' are composed of gangetic alluvium. The Trans Yamuna region is mainly quartzitic in nature and consisting denudational hills. Administratively the district comprised of eight tehils and twenty blocks. The mineral products that are commonly found in the district are glass sand, building stone, kankar, brick earth and reh. Building stone is extracted either by blasting or by splitting the chief quarries. Bricks and pottery, earth-material is available in the alluvial tract of the district and is locally used for the manufacture of bricks and earthenware.

3 Materials and Methods

Total fourteen Survey of India (SOI) topographical sheet at 1:50,000 scales were used 63—G/10, G/11, G/12, G/14, G/15, G/16, H/13, K/2, K/3, K/4, L/1, K/6, K/7, K/8, and L/5 and satellite images of Landsat TM, 1990, ETM$^+$ 2000 (http://www.usgs.gov/pubprod/aerial.html#satellite) and 2010 were used.

3.1 Classification of Satellite Data

According to Lu and Weng (2007), the major steps of image classification may include a suitable classification system, selection of suitable classification approaches, post-classification processing and accuracy assessment. These images were first geometrically and radiometrically corrected ERDAS IMAGINE 9.1

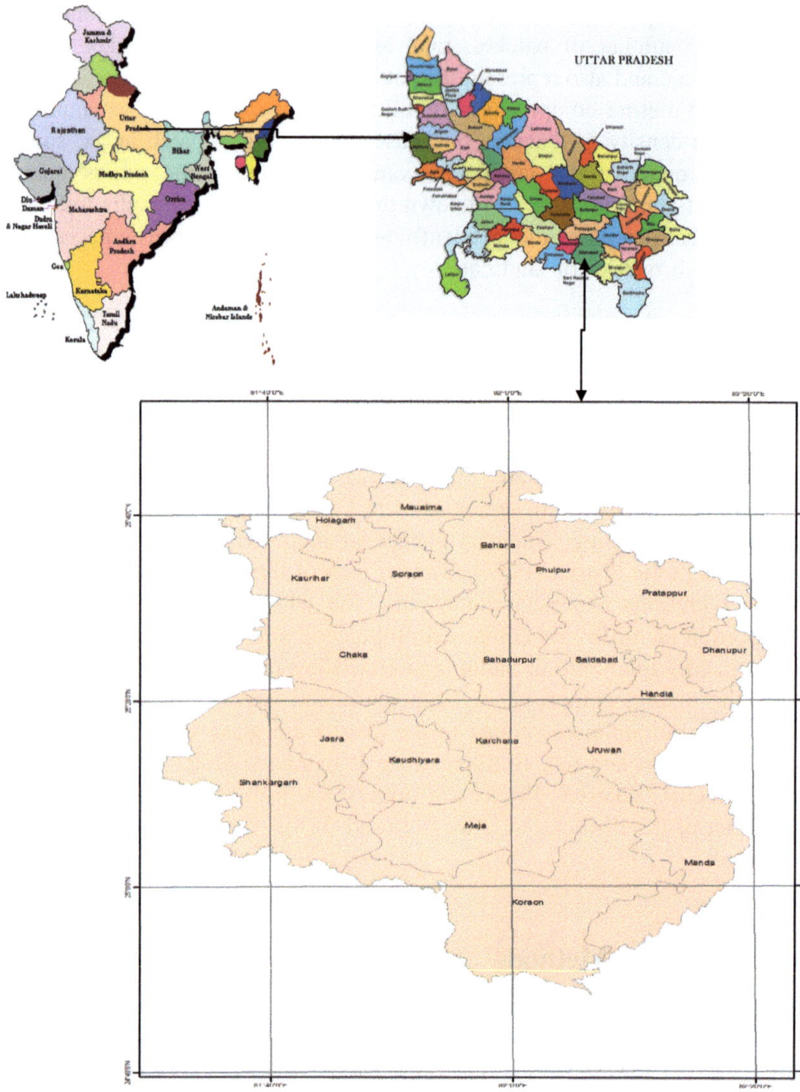

Fig. 1 Study area location map

tool (ERDAS Field Guide 1999). The automatic atmospheric calibration was performed on each image separately. The Anderson et al. (1976) classification method was used as classification scheme. Afterwards, the Maximum Likelihood Classification tool is considered for image classification of year 2010 as it is taking account of both the variances and covariances of the class signatures and assigning each cell to one of the classes represented in the signature file. The algorithm used by the Maximum Likelihood Classification tool is based on Bayes' theorem and the equation used in MLC classification as shown in the Eq. 4. The image of year 1990 and 2000 was classified using Isodata clustering approach (unsupervised method). The random sampling method was used for the accuracy assessment. The Google image, ancillary data and field data was employed for the accuracy assessment.

In the study following Eqs. 1, 2, and 3 were used e.g.:

$$\%\ Cover\ Change = \frac{Area_{i\ year\ x} - Area_{i\ year(x+1)}}{\sum_{i=1}^{n} Area_{i\ year\ x}} \times 100 \tag{1}$$

$$\%\ Cover\ Coverage = \frac{Area_{i\ year\ x}}{\sum_{i=1}^{n} Area_{i\ year\ x}} \times 100 \tag{2}$$

$$Change\ area = Area_{i\ year\ (x+1)} - Area_{i\ year\ x} \tag{3}$$

where $Area_{i\ year\ x}$ area of is cover i at the first date; $Area_{i\ year\ (x+1)}$ is area of cover i at the second date; $\sum_{i=1}^{n} Area_{i\ year\ x}$ is total cover area at the first date; t_{years} is period in years between the first and second scene acquisition dates.

$$D = \ln(a_c) - [0.5\ln(|\mathrm{cov}_c|)] - [0.5(\mathbf{X} - \mathbf{M}_c)\mathrm{T}(\mathrm{cov}_c - 1)(\mathbf{X} - \mathbf{M}_c)] \tag{4}$$

where, D is weighted distance; c is a particular class; X is the measurement vector of the particular pixel; M_c is the mean vector of the sample of class; a_c is percent probability that any particular pixel is a member of class c; (Defaults to 1.0); Cov_c is the covariance matrix of the pixels in the sample of class c; $|Cov_c|$ is determinant of Cov_c; $Cov_c - 1$ is inverse of Cov_c; ln is natural logarithm function; T = transposition function.

3.2 Fragmentation Analysis

A spatial pattern analysis program i.e., FRAGSTATS3.3 offers a comprehensive choice of landscape metrics and have been used to quantify landscape structure. It is implemented by decision maker and ecologists to analyze landscape fragmentation and to describe the characteristics and components of those landscapes. These statistics facilitates the comparison of landscapes and the evaluation of processes. The advantage of FRAGSTATS is that the calculations are implemented in a fully integrated fashion in a GIS and consequently easy to apply to digital map. The three indices have been taken into account for the ecological metrics analysis viz. patch density (PD), mean patch size (MPS) and percentage of landscape (PLAND). The number of patche per 100 ha, mean patch size (MPS) means the area occupied by a particular patch type divided by the number of patches of that type (McGarigal and Marks 1995) and percentage of landscape (PLAND) (McGarigal and Marks 1995) measure of landscape composition. The representative equations for all the three indices are shown by following Eqs. 5, 6, and 7.

$$PD = \frac{n_i}{A}(10{,}000)(100) \tag{5}$$

where n_i is number of patches in the landscape of patch type (class) i; A is total landscape area (m^2).

$$MPS = \frac{A}{N}\left(\frac{1}{10{,}000}\right) \tag{6}$$

where N is total number of patches in the landscapes, excluding any background patches

$$PLAND = \frac{\sum_{j=1}^{n} a_{ij}}{A}(100) = P_i \tag{7}$$

P_i is proportion of the landscape occupied by patch type (class) i; a_{ij} is area (m^2) of patch ij.

4 Result and Discussion

4.1 Land Use and Land Cover Distribution

The study area has been classified into different land use classes shown in Fig. 2. The overall accuracy for the classified images was arrived 80.39, 85.36 and 88.10 % respectively of years 1990, 2000, and 2010. According to Anderson et al. (1976), the accuracy of classified images should >84 % for better results. On the basis of analysis the result showed that the builtup area was 555.89 km^2 in 1990 and it increases 647.80 km^2 in 2000 and it further increases to 744.16 km^2 in 2010 respectively. This may be attributed to an increase in urban population. As the population of the district shows the increasing trend of population in the region. The area of agriculture land showed nominal increment from 3,638.40 km^2 in 1990 to 3,739.22 km^2 in 2000 and 3,793.25 km^2 in 2010, which means that the area remained same for agriculture. Cultivable area decreased from 246.19 km^2 in 1990 to 193.28 km^2 in 2000 and it further decreased to 116.97 km^2 in 2010 (Table 1). This decrease in cultivable area may be due to either urban pressure or nominal conversion into agriculture land. The difference in land use classes are given in (Table 2) which shows change in area which indicates that in the year 2000/1990

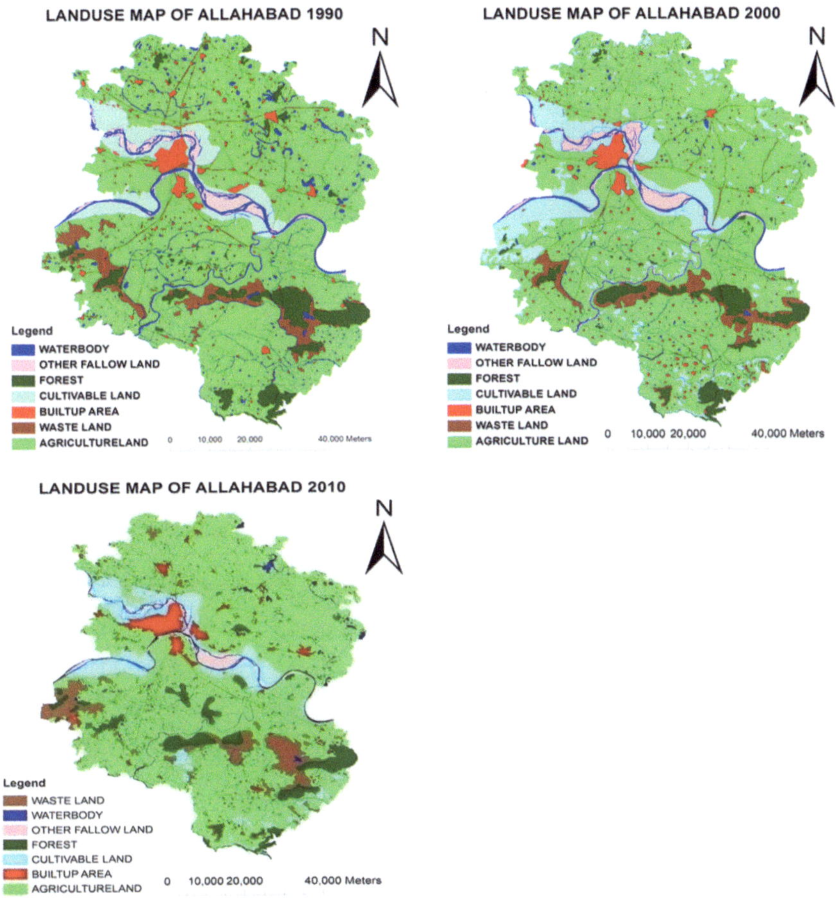

Fig. 2 Unsupervised and supervised classified satellite images of year 1990, 2000, and 2010

the forest is −86.89 km^2 while in the 2010/2000 its area is increased by 62.94 km^2, however in overall analysis it showed 2010/1990 it reduced −23.95 km^2 respectively indicates that the fluctuations in forest cover area could be due to social and economic influence. Built up area change 2000/1990 are 91.91, in 2010/2000 are 96.36 and in 2010/1990 increase in 188.3 km^2 indicates that the built up area continuously increased over the study period.

The change area of cultivable land in 2000/1990 was −52.91, in 2010/2000 was −76.31 and in 2010/1990 was −129.2 km^2. It indicates a slight decline of cultivable land due to urbanization. It indicates that this change may be due to agricultural and built-up area expansion. The agriculture area in 2000/1990 is

Table 1 Area of different land use classes in km^2 for the year 1990–2000–2010

Land use land cover distribution						
Land use land cover categories	1990		2000		2010	
	Area (km^2)	Area (%)	Area (km^2)	Area (%)	Area (km^2)	Area (%)
Forest	335.87	6.03	248.98	4.47	311.92	5.6
Waste land	279.62	5.02	234.50	4.21	179.91	3.23
Built up area	555.89	9.98	647.80	11.63	744.16	13.36
Cultivable land	246.19	4.42	193.28	3.47	116.97	2.1
Agriculture land	3,638.40	65.32	3,739.22	67.13	3,793.25	68.1
Water body	328.08	5.89	324.18	5.82	298.00	5.35
Other fallow land	186.04	3.34	182.14	3.27	125.88	2.26
Total	5,570.09	100	5,570.09	100	5,570.09	100

Table 2 Difference in area of different land use classes during 1990–2000–2010 (positive values shows decrease while negative values shows increase)

Land use classes	Land use and land covers change (area in km^2)		
	1990–2000	2000–2010	1990–2010
Forest	−86.89	62.94	−23.95
Waste land	−45.12	−54.59	−99.97
Built up area	91.91	96.36	188.3
Cultivable land	−52.91	−76.31	−129.2
Agriculture land	100.8	54.03	154.9
Water body	−3.9	−26.18	−30.08
Other fallow land	−3.9	−56.26	−60.16

100.8 km^2, while reduced in 2010/2000 is 54.3 km^2 and however it further showed increment in 2010/1990 is 154.9 km^2 respectively. Water body area in 2000/1990 is −3.9, while in 2010/2000 its area is reduced by −26.18; however in 2010/1990 again it shows reduction by −30.08 km^2 respectively, which indicates that the change in water body area could be due loss of surface water body.

4.2 Landscape Level Metrics

Landscape metrics were characterized for the landscape fragmentation patterns which reveals the configuration and composition pattern of the landscape element such as in the form of class, patch and landscape metrics. The spatial characteristics of patch, class of patches or entire landscape also quantifying and exploring with the help of ecological process by landscape metrics (Narumalani et al. 2004). The landscape level metrics analysis analyses the following parameters i.e. total area (TA), number of patches (NP), patch density (PD), and Area mean. PD metrics is an important measure to show the health of an ecosystem. If the NP increases while the area under the class decreases, it represents fragmentation or dissection. Anthropogenic and natural activities affect the spatial structure of landscape such as urbanization and industrialization due to anthropogenic activities while the natural activities like flood and drought etc. The nature of these change experienced by each land use land cover varies.

The landscape can undergo in different type of transformation, the result after the change process analysis are given in Table 3 shows that the total area 499,182.5 ha in 1990, 498,818.6 ha in 2000 and 550,046.2 ha in 2010. The number of patches per unit area has increased from 3,301 ha in 1990 to 42,394 ha in 2000 and further it decreases by 4,646 ha in 2010 respectively in twenty block of study area. The increase in number of patches in the first decade shows that the landscape is more fragmented while in the second decade it shows that the landscape gets less fragmentation. The maximum number of patches (582 in year 2010) found in Koraon block. Kaurihar and Soraon have number of patches 240, 134 in 2000 means these areas are fragmented during the year 2000. While low number of patches i.e. 54 was present in Dhanupur means this area has less developmental activities. PD increases from 12.755 in 1990 to 386.142 in 2000 and after one decade it rapidly decreases to 19.821 in 2010. These values indicate that the landscape fragmentation was at higher pace during the year from 1990 to 2000 while it was less fragmented during the year in 2000–2010. These change in values of mean area indicates that the forest is less fragmented in 2000 than in 1990 and again it is more fragmented in 2010. Two decades of natural conditions or human activities had less impact on forest landscape and thus fragmentation was less in 2000. While in 2010, human pressure or natural conditions may played a major role in the decrease of forest mean area.

4.3 Class Level Metrics

The class level metrics has a spatial feature to represent each land use land cover classes. The importance of class level metric analysis is to assess the transformation types which affect the spatial pattern of the landscape. In the present study we calculated the core area (CA), number of patches (NP), patch density (PD), and percentage of landscape (PLAND), and Area mean of land use classes. The class level metric analysis shows that CA of agriculture land (Table 4) from 345,595.3 ha in 1990 to 363,219.9 ha in 2000 which further decreases 334,817.4 ha in 2010. Out of twenty blocks the maximum core area of agricultural land was found in Koraon block 43, 684.5 in 1990, 43,683.94 in 2000 and 42,928.23 in 2010.ha, while minimum core area of agricultural land was reported in Kaundhiyera (1990) and Soraon (2010) 7,759.75 and 8,065.66 ha respectively. The overall study reveals that the year 2000 comprises the maximum core area in two decades. The number of patches of agricultural land decreases from 550 in

Table 3 Landscape level metric analyses during 1990–2000–2010

Block	Land fragmentation classes											
	TA			NP			PD			Area mean		
	1990	2000	2010	1990	2000	2010	1990	2000	2010	1990	2000	2010
Bahadur Pur (A)	19,675.54	24,451.57	24,451.85	103	105	154	0.5235	0.4294	0.6298	182.4716	232.8659	136.2284
Baheriya (B)	25,656.46	25,656.46	25,657.66	182	167	223	0.7094	0.6509	0.8691	137.2189	152.5269	113.9818
Chaka (C)	44,671.8	44,671.8	44,673.11	296	230	284	0.6626	0.5149	0.6357	10.2493	193.8678	144.8997
Dhanupur (D)	16,597.84	16,597.84	16,596.41	54	54	120	0.3253	0.3253	0.723	306.9854	307.1328	138.303
Handia (E)	16,791.73	16,791.73	16,792.56	94	107	181	0.5598	0.6372	1.0779	173.0862	154.8885	92.2627
Holagarh (F)	14,455.04	14,455.04	14,454.18	164	125	195	1.1346	0.8648	1.3491	86.3075	114.5474	182.8897
Jasra (G)	26,610.2	26,610.2	26,609	166	207	441	0.6238	0.7779	1.6573	156.5568	128.4638	59.351
Karchhana (H)	27,806.32	27,806.32	27,805.85	122	242	148	0.4387	0.8703	0.5323	212.6537	114.1218	181.7698
Kaundhiyara (I)	21,944.88	21,944.88	21,943.26	190	103	292	0.8658	0.4694	1.3307	104.3583	211.8577	69.7837
Kaurihar (J)	19,763.02	18,762.01	19,761.9	110	240	270	0.5566	0.6509	1.3663	114.1218	103.2439	72.7601
Koraon (K)	66,121.86	66,121.86	66,123.05	433	382	582	0.6549	0.5777	0.8802	152.1393	147.761	113.6001
Manda (L)	35,354.4	35,354.4	35,354.04	253	225	306	0.7156	0.6364	0.8655	138.6052	140.6936	115.4479
Mau Aima (M)	17,310.67	17,510.47	17,309.64	165	190	207	0.9532	0.3253	1.1959	104.3914	69.7837	83.6107
Meja (N)	26,610.2	46,091.37	46,090.87	198	209	359	0.6509	0.4534	0.7789	182.8897	185.3426	128.1778
Phulpur (O)	23,072.12	23,072.12	23,073.38	212	126	239	0.9189	0.5461	1.0358	108.6951	182.8897	96.5325
Pratappur (P)	21,702.91	21,641.35	21,706.54	110	100	178	0.5068	0.4621	0.82	196.8082	194.6931	121.9431
Saidabad (Q)	19,369.24	19,369.24	19,369.22	126	82	121	0.6505	0.4234	0.6247	151.0637	235.2959	157.2086
Shankargarh (R)	50,873.17	50,873.17	50,870.54	302	218	319	0.5936	0.4285	1.205	166.7587	232.7424	82.1304
Soraon (S)	14,096.19	35,354.4	14,094.56	82	134	186	0.5817	0.3253	1.3197	162.2836	103.2439	73.0113
Urwa (T)	17,309.13	17,309.29	17,308.58	137	186	160	0.7915	1.0746	0.9244	122.6177	93.0607	105.7192
Total	499,182.5	498,818.6	5,500,046	3,301	3376	4,646	12.755	386.142	19.821	2,673.25	3,229.23	2,086.722

1990 to 508 in 2000 and while it increases 799 in 2010. Out of twenty blocks NP was more in Chaka (2010), Kaundhiyera (2010) block 89 and 84 respectively. The low NP of agricultural land was found in Dhanupur (2000), Pratappur (2000) block 7 and 9 respectively, the low NP in the year 2000 shows less fragmentization of agricultural land compared to 2010. The PD was 2.18 in year 1990 of agricultral land which decrease to 1.78 in year 2000, and again increases to 3.18 in 2010. Maximum PD was found in Soraon (2010) 0.39 and Kaundhiyera (2010) 0.38 while minimum in Dhanupur and Pratappur (2000) 0.04 and 0.04 respectively. PLAND in two decade's study was found maximum in Dhanupur (1990) 93.85 and Handia (2000) 87.48, while it was minimum in Chaka (2010) 4.88 and Kaundhiyara (1990) 35.36 out of twenty block. The Area mean of agriculture land firstly increased from 14,588.2 ha in 1990 to 17,742.5 ha in 2000 and then decreased to 11,036 ha in 2010, the maximum area occupied by Dhanupur (2000) and Pratappur (2000) was 2,212 and 2,099 ha respectively. The minimum in Chaka (2010) and Kaundhiyara (2010) was 24.58 and 101.7 ha respectively.

The CA of built up area (Table 5) decreases from 25,483.8 ha in 1990 to 24,783.5 ha in 2000 and it again increases from 24,783.5 to 38,045.7 ha in 2010. Out of the twenty block maximum CA was found in Chaka (2000) 8,485.91 ha and Bahadur Pur (2010) 3490.648 ha while minimum CA was found in Bahadurpur (1990) was 270.97 and 274.7842 ha in Kaurihar (1990). The two decade's study, showed that the year 2010 has more CA 38,045.7 ha compared to 25,483.8 ha in year 1990 of builtup area. The NP increased from 741 in 1990 to 775 in 2000; it again increased to 1863 in 2010. Shankargarh (2010) and Jasra (2010) bock have more NP 240 and 212 respectively. The result indicates that the year 2010 is fragmentized as compared to previous decades. The PD has decreased from 3.4 ha in 1990 to 3.21 ha and in 2000 it further increased to 7.26 ha in 2010. PD in Holagarh (2010) 0.78 and Mau (2010) 0.56 are found at maximum but a minimum in Bahadurpur (1990) was 0.07, Urwa (2000) and Shankargarh (1990) were 0.07. The lowest PD was obsevered in Meja 0.5 in year 2000.

The PLAND also decreased from 95.07 ha in 1990 to 80.89 ha in 2000, while again increased 146.6 ha in 2010. The maximum value of PLAND was found in Soraon (2010) 15.63 ha and Chaka (2000) 18.99 ha while minimum is 1.37 ha in Bahadurpur (1990) and 1.048 ha in Meja (2000). The Area mean of built-up area increases from 689.5 ha in 1990 to 799.9 ha in 2000 and again decreases to 611.32 in 2010. Out of twenty blocks the maximum Area mean contained by Chaka 292.6 ha in 2000 and minimum in Soraon (2000) and Jasra (2010) with 6.64 and 7.55 ha respectively.

The CA for cultivable area (Table 6) decreases from 18,699.62 ha in 1990 to 14,696.69 ha in 2000 and remains to 14,696.69 ha in 2010. The maximum CA is investigated in Jasra (2000) 7,983.1 ha and Kaundhiyara (2000) 9,530.8 ha, while minimum CA was found in Dhanupur (1990) 0.0812 ha. The NP of cultivable area was found 217 in 1990 and it increases to 422 in 2000 but further it shows reduction to 311 in 2010. Out of twenty blocks maximum NP was found in Koraon (2000) 65 and minimum 1 in Bahadurpur, Dhanupur and Kaurihar (1990) respectively. The PD also decreases in two decades study period from 0.29 in 1990

Table 4 Class level metric analysis for core area, number of patch, patch density, percentage of land, and area mean (ha) during 1990–2000–2010 of agricultural land

Block	Agricultural land														
	CA			NP			PD			PLAND			Area mean		
	1990	2000	2010	1990	2000	2010	1990	2000	2010	1990	2000	2010	1990	2000	2010
Bahadur Pur	16,155.5	11,958.76	12,634.11	19	33	22	0.1	0.14	0.09	82.11	48.9079	51.6693	850.3	362.4	574.3
Baheriya	20,398.2	20,714.49	21,247.24	18	21	23	0.07	0.08	0.09	79.505	80.7379	82.8105	1,133	986.4	923.8
Chaka	22,155.6	22,183.85	2,183.255	32	50	89	0.07	0.11	0.2	49.596	49.6596	4.8872	692.4	443.7	24.53
Dhanupur	15,577.7	15,487.09	14,741	12	7	13	0.07	0.04	0.08	93.854	93.3079	88.8204	1,298	2,212	1,134
Handia	13,742.1	14,690.27	13,543.53	19	11	34	0.11	0.07	0.2	81.838	87.4852	80.652	723.3	1,335	398.3
Holagarh	11,807.4	12,470.88	11,921.49	35	13	18	0.24	0.09	0.12	81.684	86.2736	82.4778	337.4	959.3	662.3
Jasra	17,887.7	15,330.65	15,788.18	27	46	69	0.1	0.17	0.26	67.221	57.6119	59.334	662.5	333.3	228.8
Karchhana	18,965.9	18,862.48	17,300.68	28	37	29	0.1	0.13	0.1	68.207	67.8352	62.2196	677.4	509.8	596.6
Kaundhiyara	7,759.75	7,817.744	8,540.327	14	11	84	0.06	0.05	0.38	35.36	35.6244	38.92	554.3	710.7	101.7
Kaurihar	16,217.8	16,627.24	14,996.7	22	21	43	0.11	0.11	0.22	82.061	84.1331	75.8869	737.2	791.8	348.8
Koraon	43,684.5	43,683.94	42,928.23	75	68	64	0.11	0.1	0.1	66.067	66.0658	64.9217	582.5	642.4	670.8
Manda	23,611.8	22,724.64	19,595.75	45	37	46	0.13	0.1	0.13	66.786	64.2767	55.4272	524.7	614.2	426
Mau Aima	13,363.3	0.00	13,862.91	33	0.00	25	0.19	0	0.14	77.197	0.00	80.0878	404.9	476	554.5
Meja	8,860.87	27,120.54	24,476.51	0.00	27	37	0	0.06	0.08	0.00	58.8408	53.1049	576	1,004	661.5
Phulpur	17,037.9	19,516.09	18,494.75	34	17	30	0.15	0.07	0.13	73.846	84.5873	80.1562	756	1,148	616.5
Pratappur	18,974.7	18,890.17	18,457.07	10	9	13	0.05	0.04	0.06	87.429	87.0398	85.03	1,897	2,099	1,420
Saidabad	14,794.2	15,223.11	13,934.71	27	15	30	0.14	0.08	0.15	76.38	78.5943	71.9425	547.9	1,015	464.5
Shankargarh	30,585.3	29,527.64	28,814.85	58	37	45	0.11	0.07	0.09	60.121	58.0417	56.6435	527.3	798	640.3
Soraon	8,736.07	16,627.24	8,065.66	15	21	55	0.11	0.11	0.39	61.975	84.1331	57.2253	582.4	791.8	146.6
Urwa	14,139.9	13,763.09	13,290.45	27	27	30	0.16	0.16	0.17	81.638	79.5127	76.7853	523.7	509.7	443
Total	345,595.3	363,219.9	334,817.4	550	508	799	2.18	1.78	3.18	1,291.23	1,352.66	1,309.0	14,588.2	17,742.5	11,036.83

Table 5 Class level metric analysis for core area, number of patch, patch density, percentage of landscape and area mean (ha) during 1990–2000–2010 of built up area

Block	Built-up area														
	CA			NP			PD			PLAND			Area mean		
	1990	2000	2010	1990	2000	2010	1990	2000	2010	1990	2000	2010	1990	2000	2010
Bahadur Pur	270.9666	1,103.6041	3,490.648	13	22	39	0.07	0.09	0.16	1.3772	4.5134	14.276	20.84	50.16	89.5
Baheriya	1,189.459	768.7134	1,031.654	71	70	47	0.28	0.27	0.18	4.6361	2.9962	4.0208	16.75	10.98	21.95
Chaka	8,054.434	8,485.9007	6,201.948	42	29	35	0.09	0.06	0.08	18.03	18.996	13.883	191.8	292.6	177.2
Dhanupur	294.3594	528.3686	678.3259	24	21	55	0.14	0.13	0.33	1.7735	3.1834	4.0872	12.27	25.16	12.33
Handia	992.0822	732.9744	1,662.001	29	14	43	0.17	0.08	0.26	5.9082	4.3651	9.8972	34.21	52.36	38.65
Holagarh	774.5616	610.1622	1,031.766	72	71	113	0.5	0.49	0.78	5.3584	4.2211	7.1382	10.76	8.594	9.131
Jasra	877.3925	741.0157	1,600.609	32	37	212	0.12	0.14	0.8	3.2972	2.7847	6.0153	27.42	20.03	7.55
Karchhana	1,038.218	643.6269	1,153.048	22	54	44	0.08	0.19	0.16	3.7337	2.3147	4.1468	47.19	11.92	26.21
Kaundhiyara	1,178.087	1,155.4256	1,734.415	42	31	83	0.19	0.14	0.38	5.3684	5.2651	7.9041	28.05	37.27	20.9
Kaurihar	274.7842	505.1383	911.2621	14	76	118	0.07	0.38	0.6	1.3904	2.556	4.6112	19.63	6.647	7.723
Koraon	1,766.075	2,966.7431	3,347.436	92	52	153	0.14	0.08	0.23	2.6709	4.4868	5.0624	19.2	57.05	21.88
Manda	538.8467	777.6482	1,530.366	38	34	88	0.11	0.1	0.25	1.5241	2.1996	4.3287	14.18	22.87	17.39
Mau Aima	1,351.34	0.00	1,494.744	64	0.00	97	0.37	0	0.56	7.8064	0.00	8.6353	21.11	21.65	15.41
Meja	483.67	483.1263	1,315.519	0.00	23	153	0.23	0.05	0.33	0.00	1.0482	2.8542	22.76	21.01	8.598
Phulpur	1,738.621	1,283.1113	1,768.924	35	27	109	0.15	0.12	0.47	7.5356	5.5613	7.6665	23.67	47.52	16.23
Pratappur	1,035.375	861.4724	1,462.684	40	38	74	0.18	0.18	0.34	4.7707	3.9694	6.7384	25.88	22.67	19.77
Saidabad	1,179.143	1,051.7013	1,741.372	38	30	49	0.2	0.15	0.25	6.0877	5.4298	8.9904	31.03	35.06	35.54
Shankargarh	1,243.149	1241.2805	2,770.244	36	58	240	0.07	0.11	0.47	2.4436	2.44	5.4457	34.53	21.4	11.54
Soraon	1,234.62	505.1383	2,203.516	20	76	66	0.14	0.38	0.47	8.7585	2.556	15.634	61.73	6.647	33.39
Urwa	451.611	338.3834	915.2139	17	12	45	0.1	0.07	0.26	2.6091	1.9549	5.2876	26.57	28.2	20.34
Total	25,483.8	24,783.5	38,045.7	741	775	1,863	3.4	3.21	7.26	95.07	80.89	146.6	689.5	799.9	611.32

to 0.27 in 2000 and 0.22 in 2010, while the block Holagarh (2000) and Kaundhiyara (1990) has maximum 0.17 PD and Dhanupur (1990), Kaurihar (1990) and Baheriya (2010) contained minimum PD 0.01. But in two decades study period the year 1990 contained more 0.29 PD. The PLAND also decreases from 105.73 in 1990 to 41.606 in 2000 and further increases 350.22 in 2010. But the maximum value was found in Urwa (1990) 81.69 and minimum in Mau (1990) 0.001. In the two decades study reveal that PLAND was high in the year 1990 in cultivable land. The Area mean of cultivable land found in increasing order from 1,567.69 ha in 1990 to 2,037.17 ha in 2000 and it further increases from 2,037.17 to 3,312.62 ha in 2010. The maximum Area mean are 804.7 ha in Chaka (2010) and Jasra (2010) 506 ha. But the minimum Area mean are in Saidabad (1990) and Phulpur (2010) 0.045 and 0.1 ha respectively.

The CA of forest area (Table 7) shows increase from 65,311.38 ha in 1990 to 78,226.87 ha in 2000 but rapid decrease of 37,260.71 ha in 2010. Out of the twenty blocks the maximum CA are Koraon (1990) 11,982 ha and Karchhana (1990) 9,146 ha but the minimum are Mau, Phulpur and Chaka (2010) 0.223, 0.39 and 0.167 respectively. This result indicates that developmental activity and human pressure put a great impact in forest area reduction over the last two decades. The NP of forest area decreased from 250 in 1990 to 158 in 2000 and further it continues reduction to 141 in 2010. The maximum NP were found in Koraon 22 (1990) and Handia 17 (1990) and minimum in Dhanupur (2010) 1 and Saidabad 1 (1990) PD of forest area in 1990 was 1.512 and increases 4.53 in 2000, it then decreases 1.61 in 2010. The maximum PD was found in Chaka (2010) 0.98 and Urwa (2000) 0.987. The minimum PD in Saidabad 0.005 (1990) and 0.007 in Meja (2010) was obtained. The value of PLAND drastically decreases from 1,333.72 ha in 1990 to 43.10 ha in 2000 and further increases 95.39 ha in 2010. Out of twenty block the maximum PLAND 18.15 ha in Manda (1990), 18.12 ha in Koraon (1990) and 10.26 ha in Karchhana (2010) but the minimum PLAND was found at 0.016 ha in Holagarh (2000) and 0.054 ha Bahadurpur (2000). The Area mean accounts in 1990 is 8,048.71 ha which decreases to 3,131.77 ha in 2000 and goes on to increase in 2010. The maximum Area mean reported out of twenty in Jasra (1990) and Holagarh (1990) 891.5 and 788.7 ha respectively.

The CA of water body (Table 8) decrease from 34,568.5 ha in 1990 to 22,580.5 ha in 2000 and gradually increases to 37,287.5 in 2010. The CA was maximum in Chaka block (2010) 11106 ha and Koraon (1990) 3,730 ha. After two decades study revealed that the CA was found maximum in year 2010 compare to year 1990. The NP of water body was 1,071 in 1990 and it decreases 982 in 2000 and further increases 1,234 in 2010. The maximum NP was found 176 in Koraon block 1990 and minimum 7 in Holagarh in 2000. The PD in 1990 was 3.91 in 1990 and decreases 3.55 in 2000 and further increases to 4.67 in 2010. The PLAND decreases from 40.65 in 1990 to 22.66 in 2000 and increases to 31.47 in 2010. In two decades the data showed that PLAND was high in 1990 and low in 2000. But out of twenty block the value of PLAND was found higher in Meja (1990) 13.12 ha and Urwa (1990) 10.52 ha but minimum was 1.72 in Holagarh (2000), Dhanupur 0.95 in 2000. The Area mean of water body also increases from

Table 6 Class level metric analysis for core area, number of patch, patch density, percentage of landscape and area mean (ha) during 1990–2000–2010 of cultivable area

Block	Cultivable land														
	CA			NP			PD			PLAND			Area mean		
	1990	2000	2010	1990	2000	2010	1990	2000	2010	1990	2000	2010	1990	2000	2010
Bahadur Pur	0.1625	7,232.193	3,126.355	1	7	36	0.01	0.03	0.15	8	29.578	12.786	0.163	103.3	86.84
Baheriya	1,100.599	2,358.937	187.9631	9	16	3	0.04	0.06	0.01	4.29	9.1943	0.7326	122.3	147.4	62.65
Chaka	7,010.611	7,066.494	2,116	54	44	25	0.12	0.1	0.06	15.69	15.819	45.03	129.8	160.6	804.7
Dhanupur	0.0812	410.3487	0.00	1	6	0.00	0.01	0.04	0	5.08	2.4723	0.00	0.081	68.39	76.5
Handia	28.3475	603.0144	123.2864	5	10	6	0.03	0.06	0.04	0.169	3.5911	0.7342	5.67	60.3	20.55
Holagarh	0.1625	796.4924	0.00	2	24	0.00	0.01	0.17	0.00	0.001	5.5101	0.00	0.081	33.19	56.8
Jasra	3,974.014	7,983.118	5,566.202	9	20	11	0.03	0.08	0.04	14.93	30	20.919	441.6	399.2	506
Karchhana	3,136.585	5,079.812	3,839.245	16	38	10	0.06	0.14	0.04	11.28	18.269	13.807	196	133.7	383.9
Kaundhiyara	9,209.94	9,530.779	7,997.867	37	7	17	0.17	0.03	0.08	41.97	43.431	36.448	248.9	136.2	470.5
Kaurihar	0.1625	832.3938	0.5566	1	17	10	0.01	0.09	0.05	8.09	4.2119	0.0028	0.163	48.96	0.056
Koraon	1.3808	2,554.932	1,910.244	15	65	57	0.02	0.1	0.09	0.002	3.864	2.8889	0.092	39.31	33.51
Manda	0.8935	1,754.785	1,524.856	8	25	36	0.02	0.07	0.1	0.003	4.9634	4.3131	0.112	70.19	42.36
Mau Aima	0.2437	0.00	0.3896	3	0.00	5	0.02	0	0.03	0.001	0.00	0.0023	0.081	9.36	0.078
Meja	.00	1,032.695	1,945.588	0.00	20	34	0	0.04	0.07	0.00	2.2405	4.2212	0.087	51.63	57.22
Phulpur	0.4874	1,536.858	0.5009	6	16	5	0.03	0.07	0.02	0.002	6.6611	0.0022	0.97	96.05	0.1
Pratappur	0.00	1,733.666	0.4453	0.00	16	5	0	0.07	0.02	0.00	7.9882	0.0021	0.67	108.4	0.089
Saidabad	1,412.097	2,206.071	2,486.768	8	10	8	0.04	0.05	0.04	7.29	11.39	12.839	0.045	220.6	310.8
Shankargarh	1,086.384	10,847.84	1,406.189	6	32	37	0.01	0.06	0.07	2.136	21.323	2.7642	181.1	33.9	38.01
Soraon	2,061.247	832.3938	2,192.161	11	17	8	0.08	0.09	0.06	14.62	4.2119	15.553	187.4	48.96	274
Urwa	14,139.89	810.3818	703.6511	27	12	8	0.16	0.07	0.05	81.69	4.6818	4.0653	52.37	67.53	87.96
Total	18,699.62	14,696.69	14,696.69	217	422	311	0.29	0.27	0.22	105.73	41.606	350.22	4,028.99	6,497.424	6,522.351

Table 7 Class level metric analysis for core area, number of patch, patch density, percentage of landscape and area mean (ha) during 1990–2000–2010 of forest area

Block	Forest														
	CA			NP			PD			PLAND			Area mean		
	1990	2000	2010	1990	2000	2010	1990	2000	2010	1990	2000	2010	1990	2000	2010
Bahadur Pur	1,766	1,678	709.4	12	7	3	0.035	0.066	0.012	4.78	0.054	2.901	233.6	345	236.5
Baheriya	1,796	1,694	1,417	13	9	13	0.057	0.076	0.051	5.76	0.077	5.524	134.7	534.7	109
Chaka	1,895	1,754	0.167	15	10	2	0.068	0.034	0.98	6.76	0.061	2.67	430.1	345	0.084
Dhanupur	1,806	1,795	252	13	5	1	0.077	0.067	0.006	3.87	0.033	1.518	465.8	324.8	252
Handia	3,975	3,684	200.4	17	7	3	0.055	0.05	0.018	6.78	0.034	1.194	654.7	675	66.81
Holagarh	1,782	1,587	2,600	15	4	4	0.068	0.056	0.08	8.65	0.016	1.56	788.7	0.074	44.98
Jasra	1,597	1,398	2,654	16	6	7	0.046	0.034	0.05	5.56	4.088	3.45	891.5	21.45	78.97
Karchhana	9,146	8,432	2,882	13	8	12	0.054	0.076	0.043	4.87	1.45	10.36	734.7	0.195	240.1
Kaundhiyara	6,179	5,798	230.8	17	14	10	0.077	0.045	0.046	5.67	2.56	1.052	376.7	69.08	23.08
Kaurihar	882.2	7,864	2,133	12	10	15	0.061	0.065	0.076	4.464	5.45	10.79	73.52	46.15	142.2
Koraon	11,982	10,654	5,520	22	16	15	0.033	0.032	0.023	18.12	2.54	8.347	544.6	54.56	368
Manda	6,416	5,698	7,069	12	9	9	0.034	0.012	0.026	18.15	2.76	2.56	534.7	55.87	785.4
Mau Aima	1,048	9,876	0.223	4	6	3	0.023	0.56	0.017	6.055	4.76	3.67	262	46.1	0.074
Meja	1,046	8,976	6,189	13	11	3	0.567	0.734	0.007	7.67	4.87	13.43	324.8	13.99	2,063
Phulpur	1,765	1,597	0.39	10	6	2	0.043	0.875	0.009	7.65	3.78	6.45	463.7	34.1	0.195
Pratappur	34.85	126	138.1	5	4	2	0.023	0.018	0.009	0.161	0.056	0.636	6.969	0.122	69.07
Saidabad	80.33	56.87	369.2	1	6	8	0.005	0.076	0.041	0.415	0.766	1.906	80.33	25.88	46.15
Shankargarh	3,247	1,929	4,140	12	12	15	0.024	0.024	0.03	6.383	3.792	8.139	270.6	160.7	276
Soraon	2,356	1,765	0.334	13	3	4	0.086	0.65	0.028	5.08	2.86	4.87	324	134	0.084
Urwa	6,512	1,865	755.7	15	5	10	0.076	0.987	0.058	6.876	3.098	4.366	453	345	75.57
Total	65,311.38	78,226.87	37,260.71	250	158	141	1.512	4.53	1.61	133.72	43.10	1.61	8,048.71	3,231.77	4,877.26

1,345.44 ha in 1990 to 593.5 ha in 2000 and further increases 613.38 ha in 2010. Out of twenty block the maximum Area mean occupied by Chaka (2010) and Bahadurpur (2000) 137.1 and 87.23 ha respectively.

4.4 Class Level Metric Analysis

The results from class level metric analysis are discussed in this section.

Percentage of landscape (PLAND) is a unique technique which revealed landscape composition, the different landscape have different land use pattern with unique attributes like residential land use has high patch density and agriculture land use has high mean patch size (Weng 2007). The PLAND of built-up area of selected blocks are positively related to the developmental activities like commercial, transportation, industrial development and human intervention. In this study topographical factors have great impact on the change of PLAND of cultivable land. The PLAND of forest and agriculture land are found to have decreased. The PLAND of water body and built-up area also change in two decades showing in Fig. 3.

The area mean of class level metrics denote the average patch size of all patch belonged to the same land use type in landscape (Fig. 4) The area mean of urban land use type (built-up) is high compared to rural land use type (agriculture, forest and water body). The change in area mean of agriculture gradually decreases in two decades while the built up and cultivable land is positively correlated with the degree of human intervention and commercial practices. The PD is expressed by the number of patch per 100 ha. PD of any particular type of landscape denotes the current condition of all land use type metrics showing Fig. 5. The change of PD of different land use type like agriculture, built up, cultivable, forest and water body have different values compared to each year differently which revealed that the fragmentation of landscape has occurred due to human intervention and developmental activity. PD of overall twenty blocks among the five landscapes like forest area has more PD than built up, cultivable, agriculture and water body. The unusual high PD of forest area at Baheriya and Soraon block are due to road, household and commercial extension. The PD of forest and agriculture area has increased whilst the PD of built-up, cultivable and water body decreases in previous two decades gradually.

The NP provides clear information about the class level metrics high value indicates more fragmentation in particular area. The NP of agriculture and forest land is more as compared to cultivable, built-up and water body. That reaveled the change in NP which is highly correlated with the human intervention showing in Fig. 6. The core area of the twenty blocks has greatly influenced in the past two decades which reveal the continuous change of class level metrics. Among the five class showing in Fig. 7, the agriculture and cultivable land have more CA compare to built-up, forest and water body in current time.

Table 8 Class level metric analysis for core area, number of patch, patch density, percentage of landscape and area mean (ha) during 1990–2000–2010 of water body

Block	Water body														
	CA			NP			PD			PLAND			AREA MEAN		
	1990	2000	2010	1990	2000	2010	1990	2000	2010	1990	2000	2010	1990	2000	2010
Bahadur Pur	1,660	959.5	1,018	36	11	39	0.183	0.045	0.16	8.435	3.924	4.162	46.1	87.23	26.09
Baheriya	1,077	790.6	1,327	29	46	110	0.113	0.179	0.429	4.197	3.081	5.17	37.13	17.19	12.06
Chaka	3,210	2,722	11,106	34	35	81	0.076	0.078	0.181	7.186	6.092	24.86	94.42	77.76	137.1
Dhanupur	692.6	159.1	917.6	13	17	23	0.078	0.102	0.139	4.173	0.959	5.529	53.28	9.36	39.9
Handia	1,467	449.1	1,151	35	27	32	0.208	0.161	0.191	8.739	2.675	6.855	41.93	16.63	35.97
Holagarh	895.3	249.4	605.1	21	7	46	0.145	0.048	0.318	6.194	1.726	4.186	42.64	35.63	13.15
Jasra	1,932	1,906	2,213	56	54	111	0.21	0.203	0.417	7.26	7.163	8.318	34.5	35.3	19.94
Karchhana	2,474	1,458	1,727	36	46	45	0.13	0.165	0.162	8.896	5.244	6.21	68.71	31.7	38.37
Kaundhiyara	1,537	1,140	1,307	24	17	63	0.109	0.078	0.287	7.006	5.194	5.955	64.06	67.04	20.74
Kaurihar	1,675	1,077	1,527	36	51	68	0.182	0.258	0.344	8.477	5.451	7.729	46.54	21.12	22.46
Koraon	3,730	2,449	2,857	176	152	139	0.266	0.23	0.21	5.642	3.703	4.321	21.19	16.11	20.56
Manda	1,962	1,264	1,621	124	98	48	0.351	0.277	0.136	5.551	3.575	4.586	15.83	12.9	33.78
Mau Aima	982	1,344	582.1	37	79	46	0.214	0.432	0.266	5.673	2.75	3.363	26.54	13.67	12.65
Meja	1,982	1,423	1,866	12	94	100	0.033	0.204	0.217	13.1205	3.088	4.049	544.6	15.14	18.66
Phulpur	1,733	658.7	1,915	76	50	49	0.329	0.217	0.212	7.51	2.855	8.301	54.76	13.17	39.09
Pratappur	989	155.8	793.5	19	36	29	0.088	0.166	0.134	4.557	0.718	3.656	52.05	4.328	27.36
Saidabad	1,238	592.9	488.9	41	15	20	0.212	0.077	0.103	6.389	3.061	2.524	30.18	39.53	24.44
Shankargarh	2,580	1,325	2,272	155	57	97	0.305	0.112	0.191	5.072	2.604	4.467	16.65	23.24	23.43
Soraon	931.6	1,077	765.3	29	51	50	0.206	0.258	0.355	6.609	5.451	5.43	32.12	21.12	15.31
Urwa	1,821	1,380	1,228	82	39	38	0.474	0.225	0.22	10.52	7.974	7.095	22.21	35.39	32.32
Total	34,568.5	22,580.5	37,287.5	1,071	982	1,234	3.912	3.515	4.672	40.65	22.66	31.67	1,345.44	593.5	613.38

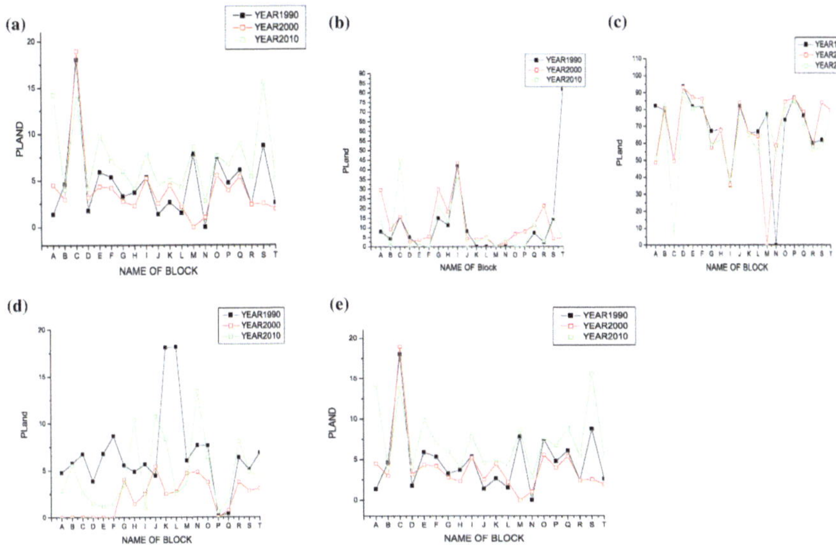

Fig. 3 Change of class level percentage of landscape (PLAND) of different land use class of **a** built-up, **b** cultivable, **c** agriculture, **d** forest and, **e** water body

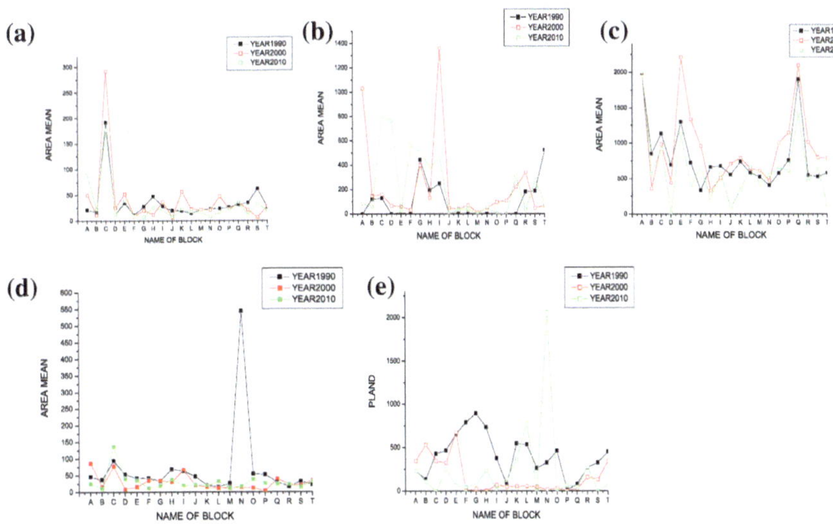

Fig. 4 Change of class level: area mean (Area Mean) of different land use class of **a** built up, **b** cultivable, **c** agriculture, **d** forest and, **e** water body

The patterns of change of class level metrics are more complicated. Generally, the results of built-up area and agriculture area are similar with metric outputs. This indicates a positive relationship to the degree of human intervention and

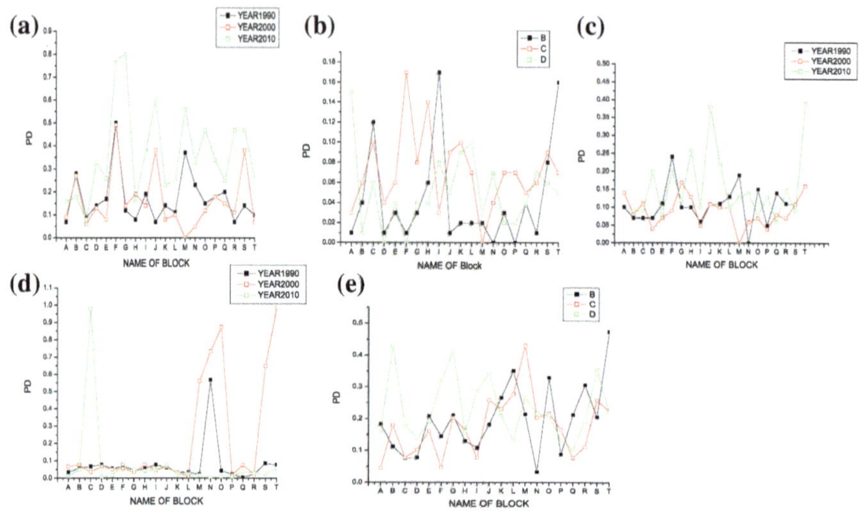

Fig. 5 Change of class level: patch density (PD) of different land use class of **a** built up, **b** cultivable, **c** agriculture, **d** forest and, **e** water body

developmental activity due to population pressure. The cultivable, forest and water body has also affected by human intervention.

4.5 Land Fragmented Class Analysis

Figure 8 shows the changes of landscape pattern both spatially and temporally through two decades. In the graph vertical axis represents the value of class and the horizontal axis represents the name of twenty blocks. TA mainly comprises of the overall area which is going through different processes of land use change. The TA value of land fragmented class when compared to previous two decades reduced and converted into many land use class with respect to the degree of human interference, as shown in Fig. 8a. The area mean shown in Fig. 8b presents gradual decrease within the last two decades indicating a fragmented area mean. Temporally, area mean in most of the blocks of land fragmented class is previously seen as decreasing and then it increases, which reveals the reciprocally changed area mean that highly correlates to the degree of human interference and developmental activity

PD at high value are indicating a fragmented landscape (Fig. 8c), which is directly or indirectly effected by the topographical or anthropogenic activity. Theoretically the observed changes of PD indicating positive relationship to the degree of commercial transportation (road construction) and residential expansions. Temporally, PD has increased through time in most of the blocks, indicating an increasing fragmentation of land fragmented class. The high value of number of

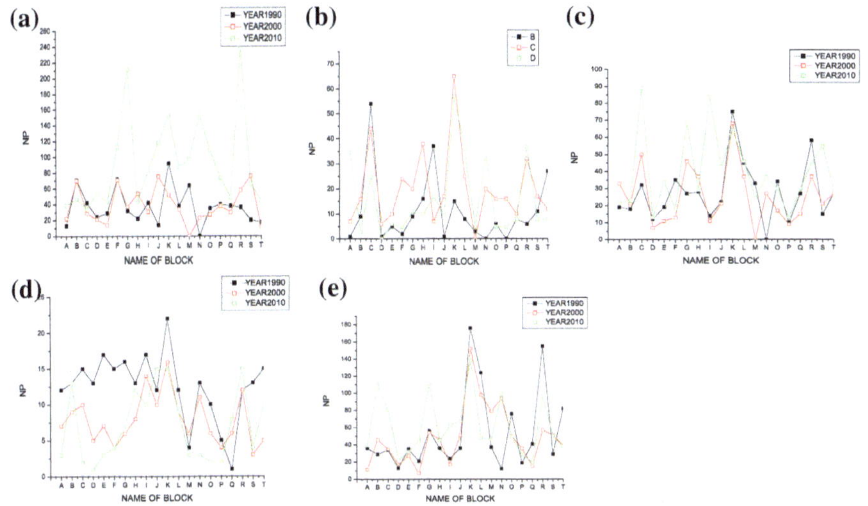

Fig. 6 Change of class level: number of patch of different land use class of **a** built up, **b** cultivable land, **c** agriculture, **d** forest and, **e** water body

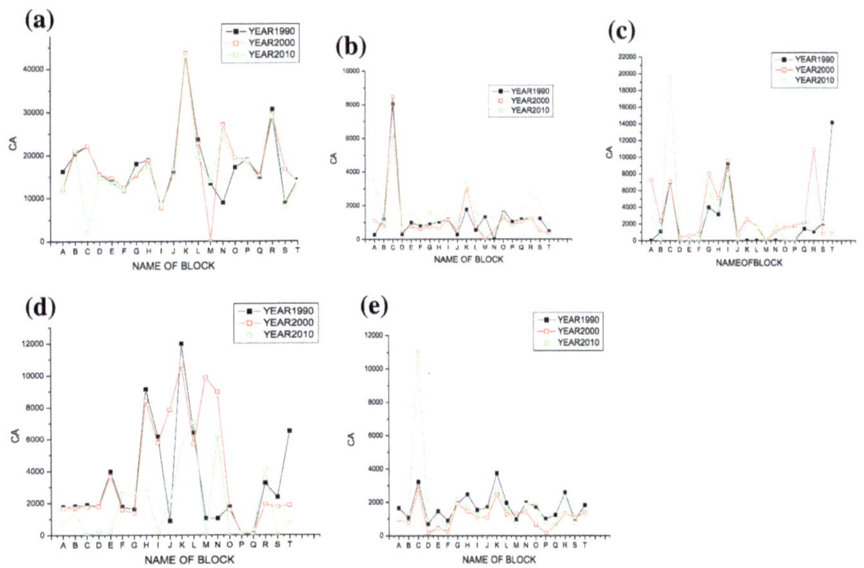

Fig. 7 Change of class level: core area of different land use class of **a** built up, **b** cultivable, **c** agriculture, **d** forest and, **e** water body

patch in a particular landscape denotes the most fragmentation. Thus the NP has significant information of current land use practice. Overall study of twenty blocks reveals that increase of NP value frequently in the current year which represents

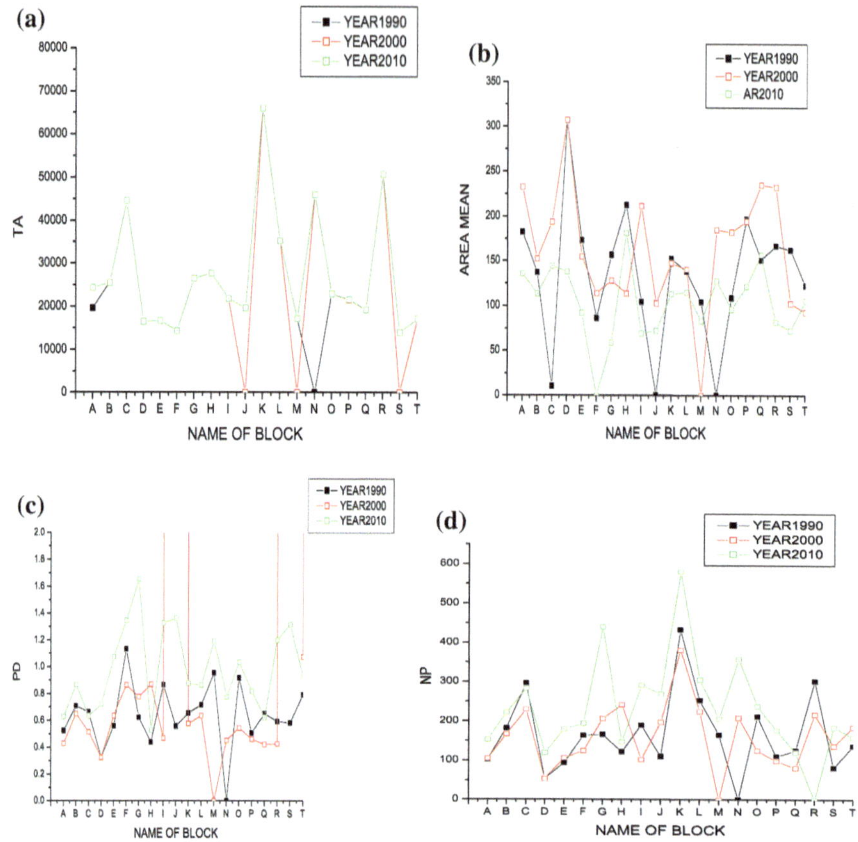

Fig. 8 **a** Total area. **b** Area mean. **c** Patch density and, **d** Number of patches in different land fragmented class

explicit information of fragmentation of landscape due to human intervention and developmental activity.

4.6 Estimating Effects on Water Quality

Rapidly growing countries, and particularly those emerging from rural to urban, frequently lack the resources to adequately monitor and forecast the impacts of land development on biologic, hydrologic, and social resources. Urbanization has significantly changed natural landscapes everywhere (Forney et al. 2001). Urban growth and fragmentation caused by urban sprawl have been extensively studied (Herold et al. 2002; Gonzalez-Abraham et al. 2007). Landscape structure is one of the most important factors influencing nutrient and organic matter runoff in watersheds (Turner et al. 2003; Wickham et al. 2003; Uuemaa et al. 2007). Therefore there is increasing demand for indicators and methods that make it possible to evaluate the landscape factors influencing water quality in freshwater management (Griffith 2002). Several studies have attempted to determine the relationship between land use/land cover structure and water quality but most studies have largely relied on compositional landscape metrics (Kearns et al. 2005). It is, however, clearly important to understand not only the total area of sources and sinks in the landscape, but also their spatial arrangement relative to flowpaths. The importance of the spatial arrangement of land cover within watersheds on water quality has been studied by Jones et al. (2001), King et al. (2005), Xiao and Ji (2007), Uuemaa et al. (2005, 2007).

The groundwater after analysis showed that the areas which are more urbanized and fragmented have elevated concentration of different ions in the water. Three sites, namely Koran, Bahariya and Karchhana, were found to have elevated fluoride in analyzed groundwater samples. Similarly, Urwa, Holagarh and Soraon had positive value of fluoride so water is not suitable for drinking. The fluoride in pre-monsoon found that Koraon, Chaka (Naini) and Bahariya had values of fluoride of 1.096–1.291 mg/l and Jasra, and Bahadurpur had moderate values of fluoride (Singh et al. 2002). The domestic pollution is the major source of pollution in Yamuna River. About 85 % of the total pollution in the river is caused by the domestic sources. About 3 km upstream from confluence with Ganga River at Grand Trunk Road Bridge called Naini Bridge near Gau Ghat (MoEF 2006).This location depicts the Yamuna River water quality before its confluence with river Ganga. The domestic pollution is mainly caused by the urban centers. The organic pollution and microbial contamination reflect increasing trend up to Allahabad. The yearly average from 1999–2005 showed that the Dissolved oxygen was highest during year 1999 and 2000 (8 mg/l) afterwards it showed decreasing trend (6.5 mg/l). During the year 2004 and 2005 Biochemical oxygen demands was higher as compared to previous years. Chemical oxygen demand at Allahabad it ranged between 5 and 18 mg/l. These results showed that the Yamuna River quality is deteriorating due to organic pollution and this may also be attributed to landscape fragmentation.

According to Singh et al. 2002, heavy metal analysis in the $<$20-μm-fraction of stream sediments appears to be an adequate method for the environmental assessment of urbanization activities on alluvial rivers. In their study they sampled during the pre-monsoon season at Delhi, Agra, Kanpur, Allahabad and Varanasi in 1993 and at Lucknow in 1995. These wide ranges are attributed to differential behavior of heavy metals rich urban effluents draining into rivers of the Ganga Plain. Stream sediments from Allahabad are classified into Sediment Pollution Index (SPI) natural sediments. To understand the impact of urbanization activities on stream sediment quality, the Pollution Load Index (PLI) given by Tomlison et al. (1980) was calculated for each urban centre. Allahabad 1.13 this PLI values show direct linear relationship with urban centre population.

5 Conclusion

The results of our study are partially validated. This study presents the results of a set of landscape metrics derived from remotely sensed data aiming to characterize the historical trends of landscape changes in the Allahabad district in the period 1990–2010. However, the identified trends in landscape changes in the region have potential policy implications in the region. The present study clearly reveals the impact and influence of the action taken during prolonged phase of land use/cover monitoring followed by a relatively short phase of passive management of twenty block of study area of Allahabad region during past two decades. The study is taken up to characterize the landscape pattern using land use/cover, landscape metrics and class level ecological metric analysis type that affect pattern of landscape changes. The landscape metric analysis indicates that from the year 1990 to 2000 periods the fragmentation of landscape was slightly low because the natural and climate condition are probably good, while from the year 2000 to 2010 it deteriorates which indicates that it could be due to human induced disturbance and development to gain more luxurious life, this still needs to be validated. The study on landscape and class level metrics helps for land management officer, forest management officer and policy makers in assessing and quantifying the extent to which the land use/cover is affecting the landscape and thus provide a complete perspective of change. Since most of changes are occurred in this area are due to human intervention and thus study over the fragmentation pattern will help to understand the spatial landscape transform processes which are also important for the ecological system and services. Landscape metric and landscape transformation analysis showed that over the time spatial configuration and composition of the landscape has changed a lot which leads to the environmental degradation. This study demonstrates the probable use of remote sensing, GIS and FRAGSTAT in assessing spatial structure and change in landscape.

Acknowledgments This research works is supported by K. Banerjee Centre of Atmospheric and Ocean Studies, IIDS, Nehru Science Centre University. Authors also thanks to the Landsat (http://www.usgs.gov/pubprod/aerial.html#satellite) programme for providing the satellite data. Authors are also thankful to the University Grant Commission, Delhi, for providing the financial grant for this research [Grant No. F. No. 42-74/2013 (SR)].

References

Anderson JR, Ernest HE, John RT, Richard WE (1976) A land use and land cover classification system for use with remote sensor data. Geological survey professional paper No. 964, U.S. Government Printing Office, Washington, DC, p 28

Central Pollution Control Board (Ministry of Environment and Forests) (2006) Water quality status of Yamuna River (1999–2005). Assessment and Development of River Basin Series: ADSORBS/41/2006-07

Dale VH, King AW, Mann LK, Washington-allen RA, Mccord RA (1998) Assessing land-use impact on natural resources. Environ Manage 22(2):203–211

ERDAS Field Guide (1999) Earth resources data analysis system. ERDAS Inc. Atlanta, Georgia, p 628
Forman RTT, Godron M (1986) Landscape Ecology. Wiley, New York
Forney W, Richards L, Adams KD, Minor TB, Rowe TG, Smith JL, Raumann CG (2001) Land use change and effects on water quality and ecosystem health in the Lake Tahoe Basin, Nevada and California. U.S. Department of the Interior U.S. Geological Survey Open-File Report 01-418
Gonzalez-Abraham CE, Radeloff VC, Hammer RB, Hawbaker TJ, Stewart SI, Clayton MK (2007) Building patterns and landscape fragmentation in northern Wisconsin, USA. Landscape Ecol 22(2):217–230. doi:10.1007/s10980-006-9016-z
Griffith JA (2002) Geographic techniques and recent applications of remote sensing to landscape-water quality studies. Water Air Soil Pollut 138(1–4):181–197. doi:10.1023/A:10155469 15924
Herold M, Scepan J, Clarke KC (2002) The use of remote sensing and landscape metrics to describe structures and changes in urban land uses. Environ Plann A 34(8):1443–1458. doi:10. 1068/a3496
Jones KB, Neale AC, Nash MS, Van Remortel RD, Wickham JD, Riitters KH, O'Neill RV (2001) Predicting nutrient and sediment loadings to streams from landscape metrics: a multiple watershed study from the United States Mid-Atlantic Region. Landscape Ecol 16(4):301–312. doi:10.1023/A:1011175013278
Kearns FR, Kelly NM, Carter JL, Resh VH (2005) A method for the use of landscape metrics in freshwater research and management. Landscape Ecol 20(1):113–125. doi:10.1007/ s10980-004-2261-0
King RS, Baker ME, Whigham DF, Weller DE, Jordan TE, Kazyak PF, Hurd MK (2005) Spatial considerations for linking watershed land cover to ecological indicators in streams. Ecol Appl 15(1):137–153. doi:10.1890/04-0481
Lal R (1998) Soil erosion impact on agronomic productivity and environment quality. Crit Rev Plant Sci 17:319–464
McGarigal K, Marks BJ (1995) FRAGSTATS: spatial pattern analysis program for quantifying landscape structure. General Technical Report PNW-GTR-351. USDA Forest Service, Pacific Northwest Research Station. Portland, OR
Morley SA, Karr JR (2002) Assessing and restoring the health of urban streams in the puget sound basin. Conserv Biol 6(16):1498–1509
Narumalani S, Mishra DR, Rothwell RG (2004) Analyzing landscape structural change using image interpretation and spatial pattern metrics. GISci Remote Sens 41(1):25–44
Pimentel D, Harvey C, Resosudarmo P, Sinclair K, Kurz D, McNair M, Crist S, Shpritz L, Fitton L, Saffouri R, Blair R (1995) Environmental and economic costs of soil erosion and conservation benefits. Science 267:2671117–2671123
Singh M, Müller G, Singh IB (2002) Heavy metals in freshly deposited stream sediments of Rivers associated with urbanization of the Ganga plain, India. Water Air Soil Pollut 141:35–54
Szabó S, Csorba P, Szilassi P (2012) Tools for landscape ecological planning–scale, and aggregation sensitivity of the contagion type landscape metric indices. Carpath J Earth Env 7(3):127–136
Tomlinson DC, Wilson JG, Harris CR, Jeffrey DW (1980) Problems in the assessment of heavy metal levels in estuaries and the formation of a pollution index. Helgel Meeresuntters 33:566–575
Turner MG (1989) Landscape ecology: the effect of pattern on process. Annu Rev Ecol 20:171–197
Turner RE, Rabalais NN, Justic D, Dortch Q (2003) Global patterns of dissolved N, P and Si in large rivers. Biogeochemistry 64(3):297–317. doi:10.1023/A:1024960007569
Uuemaa E, Roosaare J, Mander U (2005) Scale dependence of landscape metrics and their indicatory value for nutrient and organic matter losses from catchments. Ecol Ind 5(4):350–369. doi:10.1016/j.ecolind.2005.03.009

Uuemaa E, Roosaare J, Mander U (2007) Landscape metrics as indicators of river water quality at catchment scale. Nord Hydrol 38(2):125–138. doi:10.2166/nh.2007.002

Weng YC (2007) Spatiotemporal changes of landscape pattern in response to urbanization. Landscape Urban Plann 81(4):341–353. doi:10.1016/j.landurbplan.2007.01.009

Wickham JD, Wade TG, Riitters KH, O'Neill RV, Smith JH, Smith ER, Jones KB, Neale AC (2003) Upstream-to-downstream changes in nutrient export risk. Landscape Ecol 18(2):193–206. doi:10.1023/A:1024490121893

Xiao HG, Ji W (2007) Relating landscape characteristics to non-point source pollution in mine waste-located watersheds using geospatial techniques. J Environ Manage 82(1):111–119. doi:10.1016/j.jenvman.2005.12.009

Zubair AO (2006) Change detection in land use and land cover using Remote Sensing data and GIS (A case study of Ilorin and its environs in Kwara State), Department of Geography, University of Ibadan, October 2006

Chlorophyll Retrieval Using Ground Based Hyperspectral Data from a Tropical Area of India Using Regression Algorithms

M. Gupta, Prashant K. Srivastava, S. Mukherjee and G. Sandhya Kiran

Abstract Newly emerged hyperspectral techniques make it possible to acquire images in narrow and continuous spectral bands, providing significant improvements when compared with broad bands. In this work total chlorophyll, chlorophyll-a and chlorophyll-b were estimated, which have been used to find an empirical relationship with the hyperspectral data. The vegetation indices used in this study are Red Edge Inflection Point (REIP), Normalized Difference Vegetation Index (NDVI), Soil Adjustment Vegetation Index (SAVI) and Structure Insensitive Pigment Index (SIPI). These indices were calculated on the first derivative of reflectance obtained from the field. The REIP value was calculated using a four point interpolation technique. The REIP value for fertilizer plots was found higher than for control, indicating a better health of crop in the fertilizer plots. Stepwise regression analysis was employed to estimate the linear regression algorithm for chlorophyll retrieval. The analysis shows that REIP and NDVI served as best predictor for total chlorophyll and chlorophyll-b estimation, while REIP was found suitable for Chlorophyll-a estimation. A global sensitivity analysis method was developed to estimate the uncertainty associated with the proposed chlorophyll retrieval algorithms.

Keywords REIP · Continuum removal (CR) · Vegetation indices · Hyperspectral data · Linear regression algorithm

---

M. Gupta (✉)
Department of Civil Engineering, Indian Institute of Technology, New Delhi 110016, India
e-mail: manikagup@gmail.com

P. K. Srivastava
Department of Civil Engineering, University of Bristol, Bristol BS8 1TR, UK

S. Mukherjee
School of Environmental Sciences, Jawaharlal Nehru University, New Delhi 110067, India

G. Sandhya Kiran
Department of Botany, M S University of Baroda, Vadodara, Gujarat, India

P. K. Srivastava et al. (eds.), *Remote Sensing Applications in Environmental Research*, Society of Earth Scientists Series, DOI: 10.1007/978-3-319-05906-8_10,

1 Introduction

Vegetation monitoring continues to be a key topic in the science and applications of remote sensing techniques (Delegido et al. 2010). In recent years, many studies and experiments have been carried out to understand the relationship between optical remote sensing and photosynthetic pigment properties and content within plant tissues. Each of these pigments have different spectral absorption features, which allow remote sensing techniques to assess vegetation (Blackburn 1998). The spectral measurement of canopy reflectance is the most common approach for non-destructive measurements of crop canopy parameters (Behrens et al. 2006). When developing methods and algorithms based on measurements of spectral reflectance data, it is necessary to know the influence of crop parameters on their spectral properties (Larsolle and Muhammed 2007; Gupta and Srivastava 2009).

The study of chlorophyll is important because it is a key component in the study of plant fluorescence. Chlorophyll fluorescence provides one of the most novel methods in remote sensing to study plant stress associated with drought, salinity, heavy metals, cold and high temperatures (Flexas et al. 2000; Xue and Yang 2009). Chlorophyll content can be used as an index of the photosynthetic potential as well as of plant productivity (Filella et al. 1994; Carter 1998; Xue and Yang 2009; Schlerf et al. 2010).

Vegetation indices can be effectively used to evaluate the overall photosynthetic capacity or productivity of a canopy and for the application to variable rate fertilization prescriptions (Vincini et al. 2008; Miao et al. 2009). The hyperspectral indices can improve accuracy of estimation for many vegetation parameters compared with multi-spectral indices (Thenkabail et al. 2000, 2004a, b). Correlation of discrete spectral bands in the form of vegetation indices (VIs) helps in estimating crop canopy parameters (Graeff and Claupein 2003). Hyperspectral (typically with bands about 10 nm in width) reflectance data are useful in precision agriculture and for assessing the physiological status of the plant (Wu et al. 2008, 2010). In-field variability can be recorded using methods based on measurements of reflectance of field crops to estimate crop status. This can then be used as a basis for site specific operations such as application of plant nutrition or pesticide dose. With spectral sensing, sampling of reflectance spectra of a growing crop can be done relatively fast and the signal from a spectral sensor can be used to control site specific inputs either indirectly by mapping the in-field variability or in real time using a spectral sensor (Larsolle and Muhammed 2007). Several researchers demonstrated that chlorophyll concentration is related to nitrogen dose given to the plants (Cho and Skidmore 2006; Fillella and Penuelas 1994; Jongschaap and Booij 2004; Haboudane et al. 2004; Hansen and Schjoerring 2003; Mutanga and Skidmore 2007).

The red edge reflectance has been a focus of research in remote sensing of vegetation, since most of the canopy spectral information is contained in the red and near infrared bands. The position and shape of red edge are indicative of plant chlorophyll content, biomass and plant water content (Filella and Penuelas 1994).

Carter and Knapp (2001) showed that maximum difference in reflectance between healthy and stressed plants generally occurred near 700 nm. The red edge position (REP) is the point of maximum slope in the vegetation reflectance spectra (Fillella and Penuelas 1994) that occurs in the 680–750 nm region. This phenomenon is caused by strong chlorophyll absorption in the red spectrum and canopy scattering in the near infrared (Dawson and Curran 1998). An increase in chlorophyll concentration or biomass results in the broadening of the absorption feature centred around 670 nm, causing the movement of the red edge position to longer wavelengths (Dawson and Curran 1998). Red edge is less sensitive to soil background and atmospheric effects and can provide information, not available from a combination of near infrared and visible spectral bands (Clevers 1999). Experimental and theoretical studies showed that REP shifts according to changes of plant health levels (Vane and Goetz 1988; Gupta and Srivastava 2009). When a plant is healthy with high chlorophyll content and high Leaf Area Index (LAI), the REP shifts toward the longer wavelengths; when it suffers from disease or chlorosis and low LAI, it shifts toward the shorter wavelengths (Pu et al. 2003). The red edge variables usually were derived from the first derivative of the reflectance. The first derivative is commonly used to enhance absorption features that might be masked by interfering background absorption and canopy background effects (Dawson and Curran 1998). Although a number of studies have been carried out on hyperspectral data, there is negligible work being carried out on chlorophyll retrieval using linear regression algorithms. Hence, the main objectives of this study were focused on (1) Evaluating the response of the red edge measured at canopy level to a variation in the nutrient supply and (2) To develop linear regression algorithms for chlorophyll retrieval.

2 Materials and Methods

2.1 Field Experiment and Canopy Spectral Measurements

Two plots of size 4 × 4 m^2 each were designed for this experiment and were sown with winter wheat. Equal irrigation treatments were given to both the plots. A recommended fertilizer dose of 120:60:50 (NPK) was provided to one plot and the second was a control without any fertilizer dose. Field canopy reflectance measurements were acquired at the most important stage of the wheat life cycle that is first tillering stage (Zadoks et al. 1974). Vegetation radiance measurements were made over each plot, using an ASD Field Spec Pro spectrometer (Analytical Spectral Devices, Boulder, Co., USA). This spectrometer was fitted with a 25° instantaneous field of view fiber optics, operated in the 350–1,050 nm spectral region with a sampling interval of 1.5 nm between 350 and 1,050 nm. The two data sets consisted of 60 hyperspectral crop reflectance data with the corresponding measurements of all the three forms of chlorophyll in fertilizer treated and control plant respectively. Prior to each reflectance measurement, the radiance of a white standard panel coated with $BaSO_4$ with a known reflectivity was recorded for normalization of the target measurements. All canopy spectral measurements were taken from a height of 1.0 m above the canopy at nadir position under cloudless or near cloudless conditions between 10:00 and 14:00. The measurement set-up ensured that the ratio of direct to diffuse incoming solar radiation was approximately constant.

2.2 Data Analysis

Continuum removal was applied on the absorption feature centered on 670 nm. Continuum removal normalizes reflectance spectra to allow comparison of individual absorption features from a common baseline (Kokaly 2001; Gupta and Srivastava 2009). The first derivative spectrum was calculated from each reflectance spectrum. First difference transformation of the reflectance spectrum calculates the slope values from the reflectance and can be derived from the following equation (Dawson and Curran 1998):

$$FDS_{\lambda(i)} = \frac{(R_{\lambda(j+1)} - R_{\lambda(j)})}{\Delta_\lambda} \tag{1}$$

where, FDS is the first derivative reflectance at a wavelength i that is a midpoint between wavebands j and $j + 1$. $R_{\lambda(j)}$ is the reflectance at the j waveband, $R_{\lambda(j+1)}$ is the reflectance at the $j + 1$ waveband and $\Delta\lambda$ is the difference in wavelengths between j and $j + 1$.

We computed the Red Edge Inflection Point (REIP) for vegetation health monitoring, Normalized Difference Vegetation Index (NDVI) as a representative of ratio indices, Soil-Adjusted Vegetation Index (SAVI) as a representative of soil-based indices, and Structurally Insensitive Pigment Index (SIPI) as a representative of Chlorophyll based indices. To investigate the use of REIP as a means to classify vegetation, a four point interpolation (linear) technique was used. The linear approach is computationally simple, robust, and the most practical and suitable method for extracting the REIP from hyperspectral data rather than other methods because only four bands and a simple interpolation computation are needed (Clevers et al. 2002). Baret et al. (1987) have applied a simple linear model to the red infrared slope. They used four wavelength bands, centred at 670, 700, 740 and 780 nm. Reflectance measurements at 670 and 780 nm are used to estimate the inflection point reflectance (Eq. 2) and a linear interpolation procedure is applied between 700 and 740 nm to estimate the wavelength of the inflection point (Eq. 3) (Kumar et al. 2001)

$$R_{REP} = (R_{670} + R_{780})/2 \tag{2}$$

$$\lambda_{REP} = \lambda_{700} + (\lambda_{740} - \lambda_{700}) \times \left[\frac{R_{REP} - R_{700}}{R_{740} - R_{700}}\right] \tag{3}$$

The most common indices are ratio indices and soil-based indices utilizing discrete red and NIR bands where vegetation reveals distinctive reflectance properties. Ratio-based vegetation indices are often preferred to soil-based indices as the soil spectral characteristics needed to establish the soil line are often unavailable or are influenced by soil variability (Broge and Mortensen 2002). Becker and Choudhury (1988) found that the NDVI was sensitive to chlorophyll absorption. It measures the health of crop and is calculated by the formula (Tucker 1979),

$$NDVI = (R_{NIR} - R_{red})/(R_{NIR} + R_{red}) \tag{4}$$

Huete (1988) proposed another vegetation index, SAVI, intended to minimize the effects of soil background on the vegetation signal by incorporating a constant soil adjustment factor (L) into the NDVI equation. He showed that for L = 0, SAVI becomes equal to NDVI.

$$SAVI = ((L + 1)(R_{NIR} - R_{red}))/(R_{NIR} + R_{red} + L) \tag{5}$$

where R_{NIR} is the reflectance at near infra red (NIR), R_{Red} is the reflectance in the red region and L is the soil adjustment factor ranging from 0 to 1.

SIPI employs the ratios of reflectance at 800, 445 and 680 nm, and can be used to accurately estimate the ratios of carotenoids to chlorophyll-a. It minimizes the effects of radiation at the leaf surface and internal structures in the mesophyll, which allows for precise estimation of pigment concentration from reflectance spectra (Penuelas et al. 1995).

$$SIPI = (R_{800} - R_{445})/(R_{800} + R_{680}) \tag{6}$$

2.3 Chlorophyll Estimation

Approximately 0.5 g of fresh sample were weighed and ground by hand in acetone with the aid of quartz sand until no green colour was left in the residual material. After the chlorophyll solution was decanted off the residue, acetone was added to a volume of 50 ml.

The volume was centrifuged for 30 min, and then measured in a laboratory spectrophotometer (Hitachi Model, U-1500, UV-Visible Spectrophotometer). Chlorophyll-a was measured at its absorption maximum of 663 nm and chlorophyll-b at 646 nm (Arnon 1949). The equations used for determination of chlorophyll content (mg/l) are-

$$\text{Tot Chl (mg/L)} = 20.2\text{A}_{645} + 8.02\text{A}_{663} \tag{7}$$

$$\text{Chl}-\text{a (mg/L)} = 12.7\text{A}_{663} - 2.69\text{A}_{645} \tag{8}$$

$$\text{Chl}-\text{b (mg/L)} = 22.9\text{A}_{645} - 4.68\text{A}_{663} \tag{9}$$

2.4 Statistical Analysis

2.4.1 Descriptive Statistics

The mean, median, lower quartile (25 %), upper quartile (75 %), standard error and the standard deviation were generated for all the parameters. We use kurtosis and skewness functions to compare variation among the corrected and uncorrected indices generated at different pressure levels. Skewness is a measure of symmetry, or more precisely, the lack of symmetry, while Kurtosis is a measure of whether the data is peaked or flat relative to a normal distribution. For univariate data Y_1, $Y_2,\ldots, Y_N$, the formula for skewness is (Natrella 2010):

$$skewness = \frac{\sum_{i=1}^{N} (Y_i - \bar{Y})^3}{(N-1)s^3} \tag{10}$$

The skewness for a normal distribution is zero, and any symmetric data should have a skewness near zero. Negative values for the skewness indicate data that are skewed left and positive values for the skewness indicate data that are skewed right.

For univariate data Y_1, $Y_2,\ldots, Y_N$, the formula for kurtosis is Natrella (2010):

$$kurtosis = \frac{\sum_{i=1}^{N} (Y_i - \bar{Y})^4}{(N-1)s^4} \tag{11}$$

where $\bar{Y}$ is the mean, s is the standard deviation, and N is the number of data points.

Data sets with high kurtosis tend to have a distinct peak near the mean, decline rather rapidly, and have heavy tails. Data sets with low kurtosis tend to have a flat top near the mean rather than a sharp peak. Negative values of kurtosis are attained for platykurtic distributions, while positive values of kurtosis are attained for leptokurtic distributions. In this analysis, a Kolmogorov–Smirnov (K–S) distribution value is generated and statistical significance for the difference within groups established at the probability level of less than 0.001. This test tries to determine whether the two datasets differ significantly or not. The KS-test has the advantage of making no assumption about the distribution of data (technically non-parametric and distribution free). The Kolmogorov–Smirnov statistic for a given cumulative distribution function $F(x)$ can be defined as:

$$D_n = \sup_x |F_n(x) - F(x)| \tag{12}$$

where sup x is the supremum of the set of distances. If the sample comes from distribution $F(x)$, then D_n converges to 0 almost surely. The empirical distribution function F_n for n observations X_i can be obtained by using the equation:

$$F_n(x) = \frac{1}{n}\sum_{i=1}^{n} I_{X_i} \leq x \tag{13}$$

where $I_{X_i} \leq x$ is the indicator function, equal to 1 if $X_i \leq x$ and equal to 0 otherwise. In practice, the statistic requires a relatively large number of data to properly reject the null hypothesis.

2.4.2 Regression Analysis

In this chapter, multivariate stepwise regression analysis was used in which the dependent (predicted) variable is the measured chlorophyll content and the independent variable is the value derived from the hyperspectral vegetation indices. The theoretical regression model is given by the following equation (Draper and Smith 1981):

$$Y = \beta_0 + \beta_1 X + \varepsilon \tag{14}$$

where Y is the variable to be predicted, X is the variable Y is predicted from, β_0 is the intercept, β_1 is the slope of the relationship between X and Y, and ε is error. Data for the analysis are supplied by paired observations of the two variables. To estimate the accuracy of the retrieved chlorophyll with the indices, we calculated the root mean square error (RMSE) as

$$RMSE = \sqrt{\sum_{i=1}^{n} (y_i - \hat{y}_i)^2 / N} \tag{15}$$

where y_i is the estimated chlorophyll through the linear regression algorithm, $\hat{y}_i$ is the observed chlorophyll, and N is the number of samples.

2.4.3 Paired t Test

A paired sample t-test is used in this study to determine the significant difference between the mean values of the predicted and observed attributes estimated through the algorithms i.e. total chlorophyll, chlorophyll-b and chlorophyll-a. Both measurements are made on each unit in a sample, and the test is based on the paired differences between these two values. The usual null hypothesis taken is that the difference in the mean values is zero.

$$t = \frac{\overline{d}}{\sqrt{s^2/n}} \tag{16}$$

where d bar is the mean difference between two samples, s^2 is the sample variance, n is the sample size and t is a paired sample t-test with $n - 1$ degrees of freedom.

3 Results and Discussion

3.1 Descriptive Statistics of the Data

The analysis of minimum and maximum values shows a high variability in the data from their respective mean values. The analysis of median value indicates the behaviour of the central or middle data distribution, while the lower quartile (25 %) and upper quartile (75 %) shows the behaviour of the lower and upper data series. The analysis of skewness shows that most of data are skewed towards the left, by contrast Chlorophyll-b and SAVI indicates a right skewed distribution. The highest skewness was shown by Chlorophyll-b followed by REIP. The kurtosis study indicates a platykurtic distribution for NDVI, SIPI and SAVI, while the rest of the parameters show a leptokurtic distribution. The highest standard error was observed with REIP in comparison to all the parameters studied. The result of the K–S distribution shows that there exists a significant difference in the population at <0.001 level of significance for Chlorophyll-b and REIP, while all other parameters show a significance at higher levels of K–S probability (Table 1).

3.2 Continuum Removal and REIP Evaluation

The spectral reflectance obtained from the field from both control and fertilizer treated plots were averaged to one spectral curve for each field experiment. The averaging resulted in two spectral curves; one being for the fertilizer treated plot and other for the control plot (Fig. 1). Continuum removal was done on the average spectral curve obtained at 590 and 750 nm (Fig. 2). The REIP value

Table 1 Descriptive statistics of the analyzed data

Parameters	Max	Min	Median	25 %	75 %	Mean	Std dev	Std. error	Skewness	Kurtosis	K–S Dist.	K–S Prob.
Total chlorophyll	13.889	8.723	12.004	11.238	12.833	11.926	1.169	0.221	−0.723	0.573	0.133	0.223
Chlorophyll a	11.599	6.39	9.835	8.878	10.889	9.74	1.378	0.26	−0.857	0.191	0.142	0.152
Chlorophyll b	5.92	1.157	2.367	2.001	2.894	2.827	1.343	0.254	1.29	0.438	0.262	<0.001
NDVI	0.735	0.264	0.545	0.442	0.632	0.53	0.129	0.0244	−0.374	−0.651	0.0774	0.839
SAVI	0.418	0.113	0.233	0.174	0.275	0.238	0.0873	0.0165	0.609	−0.288	0.126	0.29
SIPI	0.774	0.439	0.617	0.561	0.681	0.613	0.0912	0.0172	−0.158	−0.556	0.0894	0.737
REIP	739	701	734	727	736	728.679	11.665	2.205	−1.407	0.592	0.291	<0.001

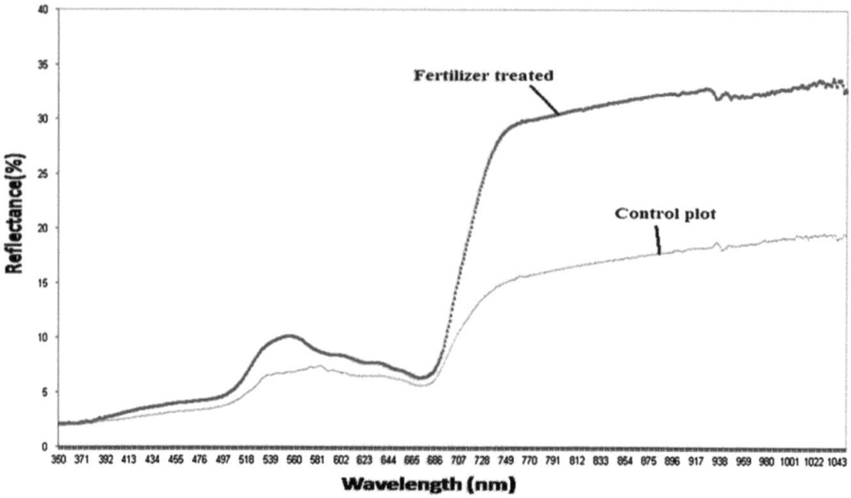

Fig. 1 Average reflectance spectra obtained from control and fertilizer treated plot

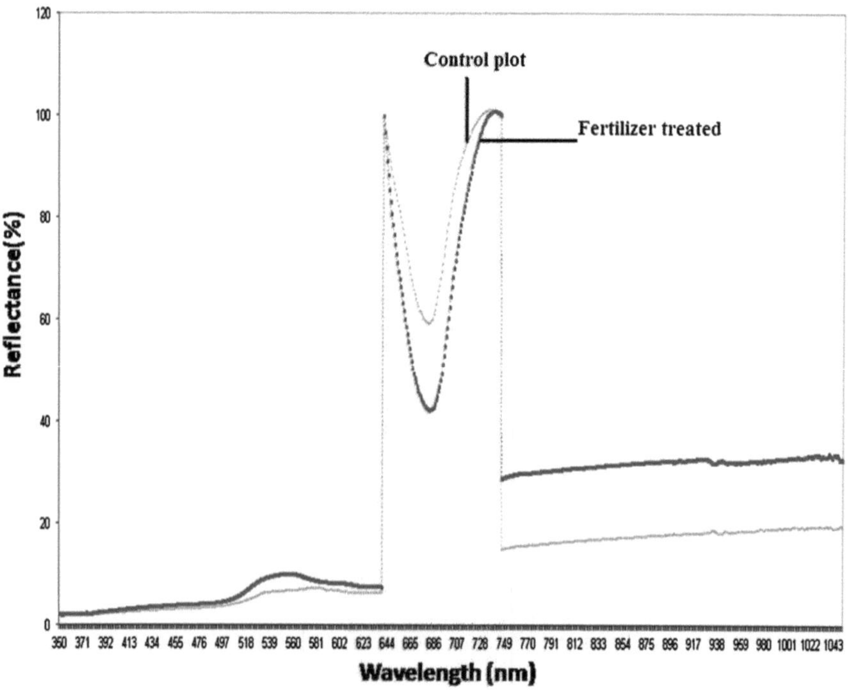

Fig. 2 Continuum removed spectra from control and fertilizer treated plot

obtained in the fertilizer treated plot indicates there is shift in the wavelength in the red region. This shift occurs due to higher amounts of nitrogen fertilizer in the treatments as compared to the control without any fertilizer dose. This indicates better health of the crop in fertilized plots while in the control plots the nutrient stressed condition is indicated.

3.3 Calculation of Vegetation Indices

The first derivative of reflectance was used to calculate different indices for quantifying the vegetation health (Fig. 3a, b). In this work, two thirds of the data are taken as a calibration dataset, while one third of the dataset is used for validation. The calibration datasets were utilized for algorithm development, while the validation dataset has been utilized for testing the algorithms. For calculation of SAVI, the value of L was assumed to be 1, 0.5 and 0 for low, moderate and high density canopy cover respectively. The canopy cover of the plots were assumed and categorised into high canopy cover (fertiliser treated) and moderate canopy cover (control plot). Hence, the value of L for fertilized treated plots was considered as 0.0, while for control plots it was computed as 0.5. Equating the values of L in the SAVI equation gives a value of SAVI equal to 0.607 (same as NDVI value) in fertiliser treated plots and 0.183 for control plots. The SIPI values were found to be 0.554 for control and 0.671 for fertiliser treated plots, indicating a stressed condition of the former compared to the latter.

3.4 Relationship of Chlorophyll with Vegetation Indices

The regression plot obtained from the stepwise regression analysis is shown in Fig. 4. In the dependent Chl-b graph the residuals are less concentrated in the centre and distributed along both tails and are skewed towards the left. In Total chlorophyll the residuals are a bit too concentrated in the center (notice the peak) which indicates a fairly a good model. In Chl-a, the concentration of residuals is little bit more than Chl-b and are skewed; notice the long tail to the right. The overall analysis of the histogram of the residual indicates that the distribution is reasonable, and can be used for implementing regression equations for predicting chlorophyll concentrations using vegetation index parameters. A probability–probability (P–P) plot showed the specified distribution of the data. For most of the data P–P plot obtained shows an approximately linear structure and hence specified distribution obtained indicates a nearly correct model. The P–P plot indicates that Chl-b shows a slightly more deviation from the mean value may be due to high variation in data and less involvement of regression components. The total chlorophyll graph indicates a more linear behaviour than others. Most of the points of the data fall within the 95 % confidence interval line. The P–P plot of Chl-a

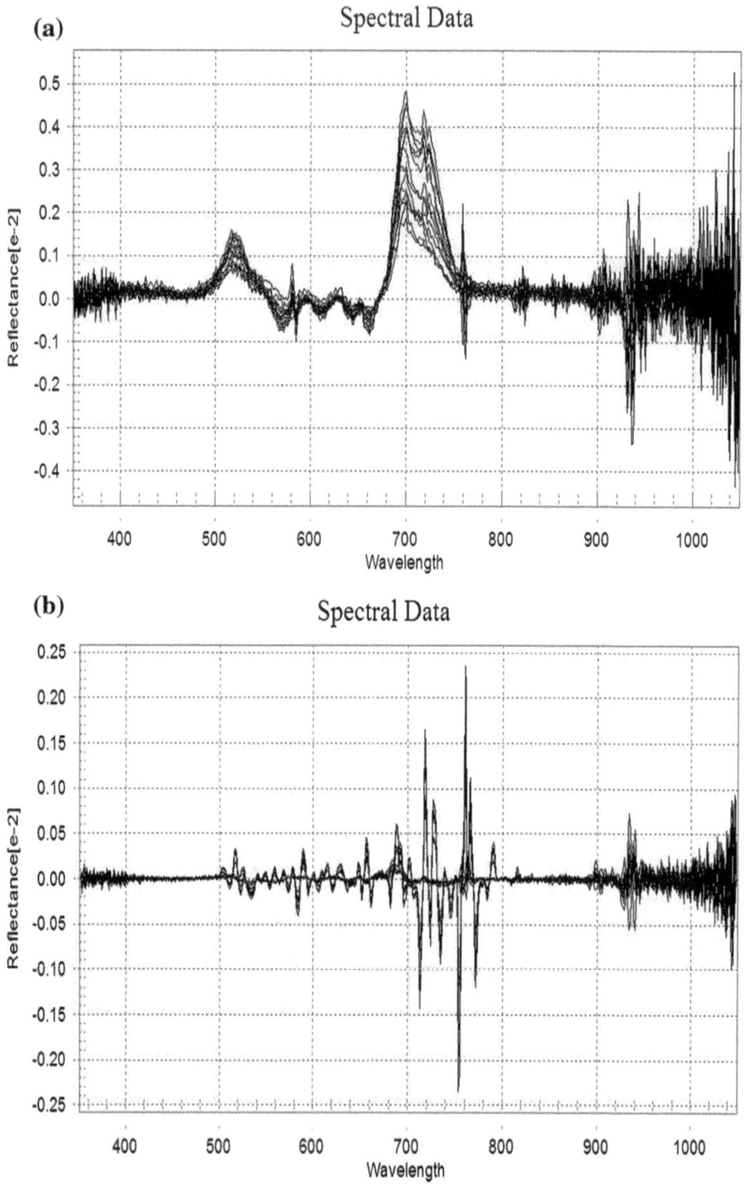

Fig. 3 **a** First derivative spectra of control plot. **b** First derivative spectra of fertilizer treated plot

indicates a linear trend, but lesser than estimated total chlorophyll which may be due to less involvement in regression statistics and may be that few points are lie outside the 95 % confidence interval. Analysis showed that all data fall within the 95 % confidence interval or prediction interval and hence data can be used for implementation of governing equations of the Chlorophyll—Vegetation indices

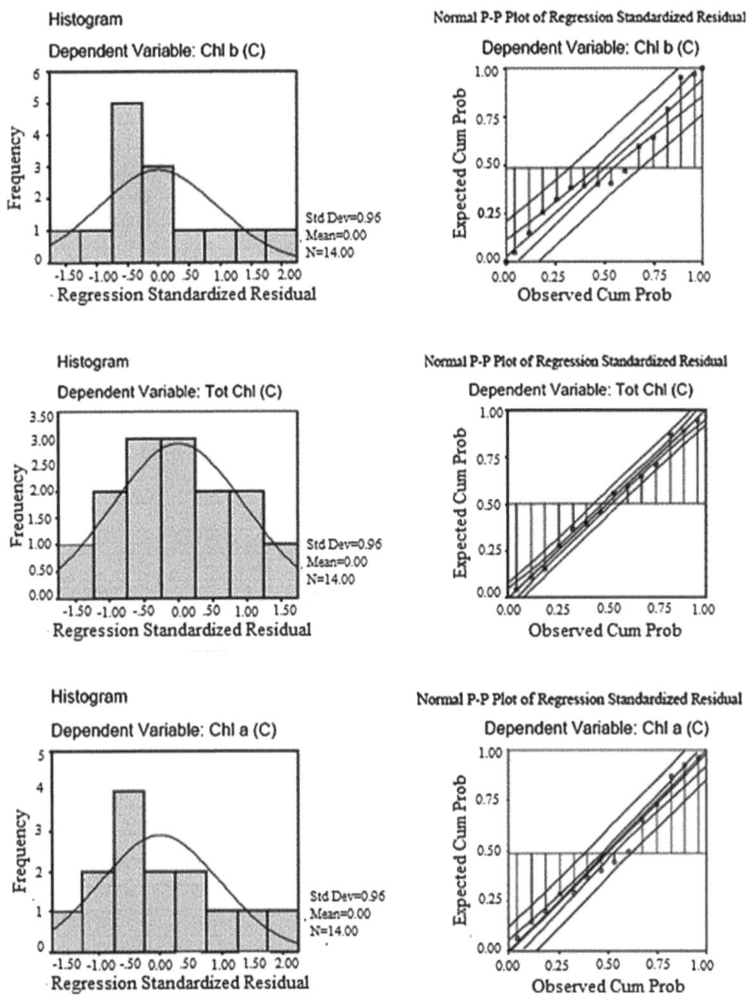

Fig. 4 Normal P–P plot of regression standardized residual between expected and observed cumulative probability with their frequency versus standardized residuals

model. Further, in all cases beta coefficients (standardized regression coefficients) are less than one which indicate there exist no multicollinearity between the data. The Coefficient of Determination (R^2) and RMSE of the chlorophyll-spectral indices model are shown in Table 2. The R^2 value was highest in Chl-a (0.706) and lowest RMSE (0.7857). For Total Chl and Chl-b, R^2 values are 0.526 and 0.520 respectively with RMSE 0.9695 and 1.0907. Vegetation indices (NDVI, SAVI, SIPI and REIP) measured from spectral data were tested to fit the multivariate regression model. The analysis shows that NDVI and REIP are the significant regression components, while SAVI and SIPI (although included in the regression model) made no significant contribution to R^2. NDVI and REIP (model 1 and 2)

Table 2 Multivariate stepwise regression analysis with their beta coefficients and t statistics

Model		Unstandardized coefficients B	Std. error	Standardized coefficients beta	t	Sig.
1[a]	(Constant)	−15.900	14.230		−1.117	0.288
	NDVI	6.647	2.351	0.592	2.828	0.016
	REIP	3.328E−02	0.020	0.351	1.678	0.121
2[b]	(Constant)	−11.454	16.010		−0.715	0.489
	NDVI	8.673	2.645	0.691	3.280	0.007
	REIP	1.451E−02	0.022	0.137	0.650	0.529
3[c]	(Constant)	−52.874	11.517		−4.591	0.001
	REIP	8.564E−02	0.016	0.840	5.371	0.000

Dependent Variables: [a] Tot Chl; [b] Chl-b; [c] Chl-a; Independent variables are REIP, NDVI, SAVI and SIPI (*Criteria* Probability-of-F-to-enter <=0.050, Probability-of-F-to-remove >=0.100). n = 40

Table 3 Equations for the conversion of the vegetation indices arbitrary values to chlorophyll content (micrograms per square centimeter)

Pigment	Linear regression algorithms	R^2	RMSE
Total Chl	Total Chl = (−15.9 + REIP * 0.033 + NDVI * 6.647)	0.526	0.9695
Chl-a	Chl-a = (−52.874 + REIP * 0.0856)	0.706	0.7857
Chl-b	Chl-b = (−11.454 + REIP * 0.015 + NDVI * 8.673)	0.520	1.0907

are the best predictors of Total chlorophyll and Chl-b estimation and account for the highest proportion of the overall variance explained, whereas for Chl-a, REIP was found moderately suitable. The various developed algorithms can be seen in Table 3. In order to test whether these regression models were reliable and applicable for Total chlorophyll, Chl-a and Chl-b estimation, the independent data sets from the validation plots were used to test the performance of the models. The result from the model was tested through a paired sample t test to know the significance of predicted values obtained from observed values. The calculated differences between these variables were normally distributed and 95 % of the differences were expected to lie between $\mu + 1.96\sigma$ and $\mu - 1.96\sigma$, known as 95 % limits of agreement. The 95 % limits of agreement for comparison of Chl-a determined with laboratory test and the Chlorophyll—spectral indices model were calculated. The results obtained for Chl-a, Chl-b and total chlorophyll are (−0.310 and 0.888); (−0.490 and 0.058) and (−0.153 and 0.820) respectively. This test shows that there exists no significant difference in the mean value of observed and predicted concentrations and hence all algorithms can be used for prediction. The standard error of mean analysis shows that the chlorophyll concentrated predicted by the model may be slightly lower or higher than the chlorophyll measured by the laboratory test (Table 4).

Table 4 Report of paired sample t test indicating calculated mean, standard deviation (SD), standard error of mean (SEM), t statistics and probability at 95 % interval

Pair(s)	Mean	SD	SEM	t statistics	Prob > t
Chl a (obs)–Chl a (Pred)	10.508	0.85699	0.22904	1.04158	0.31659
	10.21931	0.30489	0.08149		
Total Chl a (obs)–Total Chl a (Pred)	12.73179	0.60049	0.16049	1.4791	0.16293
	12.39846	0.57617	0.15399		
Chl b (obs)–Chl b (Pred)	2.72418	1.23958	0.33129	−1.70256	0.11243
	2.94018	1.22175	0.32653		

Null hypothesis mean1 − mean2 = 0; *Alternative hypothesis* mean1 − mean2 ⋦ 0; n = 20

Table 5 Sensitivity analysis of the algorithms with unperturbed and perturbed values

(R^2 and RMSE unperturbed: Total Chl = 0.53; 0.97 Chl a= 0.70; 0.79 Chl b = 0.52; 1.09)

Algorithms	Perturbation (%)		Perturbed R^2	R^2 change	Perturbed RMSE	RMSE Change
	NDVI	REIP				
Algorithm 1 (total chlorophyll)	+26.80	+2.02	0.43	−0.10	1.12	+0.15
	−26.80	−2.02	0.47	−0.06	0.94	-0.03
Algorithm 2 (chlorophyll a)	–	+2.02	0.70	0.00	0.72	−0.07
	–	−2.02	0.70	0.00	0.75	−0.04
Algorithm 3 (chlorophyll b)	+26.80	+2.02	0.38	−0.14	1.45	+0.36
	−26.80	−2.02	0.41	−0.11	1.30	+0.21

3.5 Sensitivity Analysis of the Empirical Relationships Developed

Sensitivity analysis of the output parameter was an important operation for testing the model generated by the linear regression algorithms. Since, many of the vegetation indices are spatially invariable as inferred through regression analysis (less contribution towards regression components). Hence, there may be chances of duplication and hence it can increase the probability of misjudgment. To rectify this problem a global sensitivity analysis method was developed, modified from the One Factor at Time (OAT) Method (detailed methodology was given in Morris 1991). The coefficient of variation statistical method was employed for deciding the percentage limits of perturbation for input variables shown in Table 5. The individual input layer was perturbed by changing their values as calculated through coefficient of variation (C.V.) analysis. MATLAB version 10.0 was used for the multiple regression analysis and estimation of R^2 value. To test the algorithm, the equation developed was run repeatedly by changing the input variables at a time by fixed percentage perturbation values. To know the sensitivity of the perturbed layer, the R^2 value found during the stepwise regression analysis was compared with the new R^2 value obtained from perturbed indices. The result of the sensitivity analysis was shown in the Table 5. The result showed that algorithm 2 is the least sensitive with perturbation and hence can be used a best predictor equation for Chlorophyll-a. The maximum sensitiveness was shown by the Chlorophyll-b algorithm with highest differences from unperturbed values, however, can be used for Chl-b determination with some error. The moderate sensitivity was shown by total chlorophyll algorithm with perturbation and hence it can be used for total chlorophyll prediction.

4 Conclusion

The method proposed in the present work demonstrates that the exploitation of these continuous functions can yield information on plant biochemical variables such as chlorophyll a + b content. In addition, the approach described in this chapter provides a practical method for estimating these variables by means of remote sensing techniques. This study can be applied to predict crop quality in the field. We have observed that in the spectral interval of interest, between 500 and 750 nm for the calculation of chlorophyll, the red edge parameters, REIP and NDVI, were found to be effective in measuring chlorophyll content. Based on the experimental data, we have seen a good retrieval of chlorophyll using the algorithms developed through this work. A method is thus afforded that allows us to obtain chlorophyll concentration from hyperspectral data without doing any destructive analysis. Thus, these results would have an important implication for agriculture crop status assessment.

Acknowledgment Authors are highly thankful to the University Grant Commission (UGC), India for providing financial support. Authors are also greatful to Space Application Centre, Indian Space Research Organization, Ahmedabad, India for providing technical support and information.

References

Arnon DI (1949) Copper enzymes in isolated chloroplasts. Polyphenoloxidase in Beta vulgaris. Plant physiol 24(1):1–15

Baret F, Champion I, Guyot G, Podaire A (1987) Monitoring wheat canopies with a high spectral resolution radiometer. Remote Sens Environ 22(3):367–378

Becker F, Choudhury BJ (1988) Relative sensitivity of normalized vegetation index (NDVI) and microwave polarization difference index (MPDI) for vegetation and desertification monitoring. Remote Sens Environ 24(2):297–311

Behrens T, Muller J, Diepenbrock W (2006) Utilization of canopy reflectance to predict properties of oilseed rape (*Brassica napus* L.) and barley (*Hordeum vulgare* L.) during ontogenesis. Eur J Agron 25(4):345–355

Blackburn GA (1998) Spectral indices for estimating photosynthetic pigment concentrations: a test using senescent tree leaves. Int J Remote Sens 19(4):657–675

Broge NH, Mortensen JV (2002) Deriving green crop area index and canopy chlorophyll density of winter wheat from spectral reflectance data. Remote Sens Environ 81(1):45–57

Carter GA (1998) Reflectance wavebands and indices for remote estimation of photosynthesis and stomatal conductance in pine canopies. Remote Sens of Environ 63(1):61–72

Carter GA, Knapp AK (2001) Leaf optical properties in higher plants: linking spectral characteristics to stress and chlorophyll concentration. Am J Bot 88(4):677–684

Cho MA, Skidmore AK (2006) A new technique for extracting the red edge position from hyperspectral data: the linear interpolation method. Remote Sens Environ 101(2):181–193

Clevers JGPW (1999) The use of imaging spectrometry for agricultural applications. ISPRS J Photogrammetry Remote Sens 54(5):299–304

Clevers JGPW, De Jong SM, Epema GF, Van Der Meer FD, Bakker WH, Skidmore AK, Scholte KH (2002) Derivation of the red edge index using the MERIS standard band setting. Int J Remote Sens 23(16):3169–3184

Dawson TP, Curran PJ (1998) A new technique for interpolating the reflectance red edge position. Int J Remote Sens 19(11):2133–2139

Delegido J, Alonso L, Gonzalez G, Moreno J (2010) Estimating chlorophyll content of crops from hyperspectral data using a normalized area over reflectance curve (NAOC). Int J Appl Earth Obs Geoinf 12:165–174

Draper NR, Smith H (1981) Applied regression analysis, 2nd edn. Wiley, New York

Filella I, Penuelas J (1994) The red edge position and shape as indicators of plant chlorophyll content, biomass and hydric status. Int J Remote Sens 15(7):1459–1470

Flexas J, Briantais JM, Cerovic Z, Medrano H, Moya I (2000) Steady-state and maximum chlorophyll fluorescence responses to water stress in grapevine leaves: a new remote sensing system. Remote Sens Environ 73(3):283–297

Graeff S, Claupein W (2003) Quantifying nitrogen status of corn (*Zea mays* L.) in the field by reflectance measurements. Eur J Agron 19(4):611–618

Gupta M, Srivastava PK (2009) Measuring winter wheat cultivar (*Triticum aestivum* L.) health status using hyperspectral reflectance data. In: Proceedings of ISPRS archives XXXVIII-8/W3 workshop : impact of climate change on agriculture, 17–18 Dec 2009

Haboudane D, Miller JR, Pattey E, Zarco-Tejada PJ, Strachan IB (2004) Hyperspectral vegetation indices and novel algorithms for predicting green LAI of crop canopies: modelling and validation in the context of precision agriculture. Remote Sens Environ 90(3):337–352

Hansen PM, Schjoerring JK (2003) Reflectance measurement of canopy biomass and nitrogen status in wheat crops using normalized difference vegetation indices and partial least squares regression. Remote Sens Environ 86(4):542–553

Huete AR (1988) A soil adjusted vegetation index (SAVI). Remote Sens Environ 25(3):53–70

Jongschaap REE, Booij R (2004) Spectral measurements at different spatial scales in potato: relating leaf, plant and canopy nitrogen status. Int J Earth Obs Geoinf 5(3):204–218

Kokaly RF (2001) Investigating a physical basis for spectroscopic estimates of leaf nitrogen concentration. Remote Sens Environ 75(2):153–161

Kumar L, Schmidt KS, Dury S, Skidmore AK (2001) Imaging spectrometry and vegetation science. In: van de Meer F, de Jong SM (eds) Imaging spectrometry. Kluwer Academic Press, Dordrecht, pp 111–155

Larsolle A, Muhammed HH (2007) Measuring crop status using multivariate analysis of hyperspectral field reflectance with application to disease severity and plant density. Precis Agric 8:37–47

Miao Y, Mulla DJ, Randall GW, Vetsch JA, Vintila R (2009) Combining chlorophyll meter readings and high spatial resolution remote sensing images for in-season site-specific nitrogen management of corn. Precis Agric 10:45–62

Morris MD (1991) Factorial sampling plans for preliminary computational experiments. Technometrics 33:161–174

Mutanga O, Skidmore AK (2007) Red edge shift and biochemical content in grass canopies. ISPRS J Photogrammetry Remote Sens 62:34–42

Natrella M (2010) NIST/SEMATECH e-handbook of statistical methods. http://www.itl.nist.gov/div898/handbook/
Penuelas J, Baret F, Filella I (1995) Semi-empirical indices to assess carotenoids/chlorophyll a ratio from leaf spectral reflectance. Photosynthetica 31(2):221–230
Pu R, Gong P, Biging G, Larrieu MR (2003) Extraction of red edge optical parameters from hyperion data for estimation of forest leaf area index. IEEE Trans Geosci Remote Sens 41(4):916–921
Ruiz-Espinoza FH, Murillo-Amador B, Garcia-Hernandez JL, Fenech-Larios L, Rueda-Puente EO, Troyo-Dieguez E, Kaya C, Beltran-Morales A (2010) Field evaluation of the relationship between chlorophyll content in basil leaves and a portable chlorophyll meter (spad-502) readings. J Plant Nutr 33:423–438
Schlerf M, Atzberger C, Hill J, Buddenbaum H, Werner W, Schuler G (2010) Retrieval of chlorophyll and nitrogen in Norway spruce (Picea abies L. Karst.) using imaging spectroscopy. Int J Appl Earth Obs Geoinf 12:17–26
Thenkabail PS, Smith RB, De Pauw E (2000) Hyperspectral vegetation indices and their relationships with agricultural crop characteristics. Remote Sens Environ 71(2):158–182
Thenkabail PS, Enclona EA, Ashton MS, Legg C, De Dieu MJ (2004a) Hyperion, IKONOS, ALI, and ETM+ sensors in the study of African rainforests. Remote Sens Environ 90:23–24
Thenkabail PS, Enclona EA, Ashton MS, Van Der Meer B (2004b) Accuracy assessments of hyperspectral waveband performance for vegetation analysis applications. Remote Sens Environ 91:354–376
Tucker CJ (1979) Red and photographic infrared linear combinations for monitoring vegetation. Remote Sens Environ 8(2):127–150
Vane G, Goetz AFH (1988) Terrestrial imaging spectroscopy. Remote Sens Environ 24(1):1–29
Vincini M, Frazzi E, D'Alessio P (2008) A broad-band leaf chlorophyll vegetation index at the canopy scale. Precis Agric 9:303–319
Wu C, Niu Z, Tang Q, Huang W (2008) Estimating chlorophyll content from hyperspectral vegetation indices: modeling and validation. Agric For Meteorol 148(8–9):1230–1241
Wu C, Han X, Niu Z, Dong J (2010) An evaluation of EO-1 hyperspectral hyperion data for chlorophyll content and leaf area index estimation. Int J Remote Sens 31(4):1079–1086
Xue L, Yang L (2009) Deriving leaf chlorophyll content of green-leafy vegetables from hyperspectral reflectance. ISPRS J Photogrammetry Remote Sens 64:97–106
Zadoks JC, Chang TT, Konzak CF (1974) A decimal code for growth stages of cereals. Weed Res 14(6):415–421

Remote Sensing Based Identification of Painted Rock Shelter Sites: Appraisal Using Advanced Wide Field Sensor, Neural Network and Field Observations

Ruman Banerjee and Prashant K. Srivastava

Abstract Recent advances in remote sensing can map the lithological and geological parameters in a synoptic way and hence opens up new dimensions in archaeological research. This work delineates accurate mappings of sandstone located and documented in the form of a suite of prehistoric rock-shelter sites in the Mirzapur district of Central India. Artificial Neural Network (ANN) and Maximum Likelihood Classification (MLC) techniques have been used to identify, classify and map the region under study using IRS-P6 Advanced Wide Field Sensor (AWiFS). Interpretation of data processing revealed that ANN performed better than MLC for mapping sandstone in and around the area of Mirzapur. A conspicuous pattern has been detected where the painted sandstone shelters followed the natural sandstone or host-rock formations revealing the painting activity. This demonstrates prehistoric social choice in terms of the production and consumption of rock art and the importance of local geology that governs this activity.

Keywords Remote sensing · GIS · Sandstone · AWiFS · ANN · MLC · Archaeological sites · Central India · Rock art

1 Introduction

Historically this region of Central India has played an important part delineating the wealth of India's cultural and natural resources (Banerjee and Srivastava 2013). Evidences of rock art and archaeology were first discovered as long as 120 years ago by the archaeologists A. C. Carlleyle and John Cockburn in the Indian sub-continent

R. Banerjee (✉)
Department of Archaeology and Anthropology, University of Bristol,
Bristol BS8 1UU, UK
e-mail: arxrb@bristol.ac.uk

P. K. Srivastava
Department of Civil Engineering, University of Bristol, Bristol BS8 1TR, UK

P. K. Srivastava et al. (eds.), *Remote Sensing Applications in Environmental Research*, Society of Earth Scientists Series, DOI: 10.1007/978-3-319-05906-8_11,

(Brown 1889; Cockburn 1894; Allchin and Allchin 1982). After the discovery, generations of archaeologists, anthropologists and other researchers discovered, documented various archaeological rock-shelter sites made of sandstone in this region (Tiwari 2000; Misra 2001; Pratap and Kumar 2009). Rock art research in India is shaped to its present day form through various stages. Immediately after the independence, several researchers found new sites and tried to analyse the painting in a descriptive manner without building any rigorous falsifying or confirmatory model. Next, the formative period also includes the emergence of several new painted shelters that had been properly documented, catalogued and sometimes excavated for hypothesis testing to identify the regional problem in rock-shelter archaeology. Later many researchers tried to date the paintings to assign an absolute chronology to the art of Central India with relative success. Finally, the landscape approach is incorporated by contemporary researchers to understand the locational, regional and micro-regional variations and the importance of model building in Indian rock-shelter archaeology to decode the prehistoric past through remote sensing and geographic information system applications.

The Advanced Wide Field Sensors (AWiFS) inbuilt in these satellites, map land use, land cover and vegetation. Two parallel radiometers and several channels serve as the primary equipments having brushbroom scanning technique with 740 km swath. This satellite cycles the earth within five days (Kandrika and Roy 2008; Shukla et al. 2010; Punia et al. 2011). The Cartosat data is used in this research to construct the digital elevation model of the region under purview. The elevation model accurately map the region and represent the landscape under question vividly (Yu et al. 2008; Jalan and Sokhi 2012; Lasaponara and Masini 2012). Considered together in today's world, remote sensing and geographical information system are well optimized tools to map any given features on the surface of the earth (Srivastava et al. 2012a, b). Particularly, the archaeological paradigms are directly related to the data acquisition, spatial, aspatial, temporal, visual, processual, simulation and modelling attributes of Remote Sensing and Geographical Information System (Pappu et al. 2010a, b). For the sake of brevity the entire region has been classified using two different techniques namely the Artificial Neural Network (ANN) and Maximum Likelihood Classification (MLC) (Diao et al. 2007; Gasparini et al. 2010) to evaluate their applications in prehistoric archaeology. MLC is a traditional, parametric and supervised classifier relies on the normal multivariate probability and class based representative pixilation of every image; on the contrary ANN is a non-parametric and unsupervised classifier and is supposed to generate better results than MLC (Dixon and Candade 2008; Oommen et al. 2008; Kavzoglu and Colkesen 2009).

The importance of decision making in terms of the selectivity of important materials to represent technological knowhow is well represented by Courty et al. in (2012) and this paradigm is equally relevant to understand the rock art of Central India. The sophisticated data, generated by the satellites and subsequently processed in the state of the art computer laboratory has showed high promise in the present decade in terms of vegetation and unique features mapping and monitoring (Jaiswal et al. 1999; Mukherjee 2004). Archaeological sites and survey

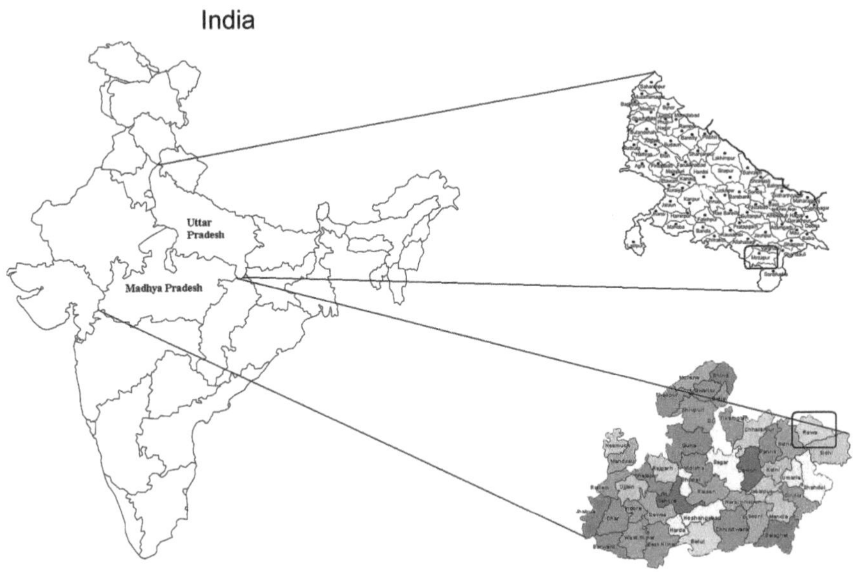

Fig. 1 Location of study area (Mirzapur and Rewa districts, India)

areas included in this research work are directly related to elevation, ranked rivulets, streams and soil types (Pappu et al. 2010a). There are very few studies reported in the Indian sub-continent, where remote sensing techniques have significantly contributed to improve field investigations, planning and management in archaeology. However, studies regarding different geographical areas are yet to be elucidated. Protection of these tangible cultural heritages now has been of utmost priority as declared by UNESCO conventions. Hence, the primary objective of this research work is; to demonstrate the results of the application of two different computational algorithms on ResourceSat-1 images to map prehistoric sandstone rock-shelters located in Mirzapur and parts of Rewa district to constrain the progressive and linear distributional trend of painted sandstone rock-shelters for predictive modelling and future archaeological surveys.

2 Material and Methodology

2.1 Study Area

Rewa (24°19′–25°12′30″ N and 81°.02′–82°.19′ E) and Mirzapur districts (23°52′ and 25°32′ N, 82°07′ and 83°33′ E), located in the state of Uttar Pradesh and Madhya Pradesh, India are well known for the archaeological and heritage value. The geographical location of the districts of Mirzapur and Rewa is shown in (Fig. 1). Both the districts are located in Central India; and specifically the study area of these two districts fall under the Vindhyan supergroup. Mirzapur district of U.P (Uttar Pradesh) shares border with the districts of Sant Ravidas Nagar and Varanasi towards the North; on the East and South-East by Chandauli and Rewa districts; on the Southern margin by Sonbhadra; and finally on the North-West by the Allahabad district of U.P. (Drake-Brockman 1911). It has an area of 4,522 km^2 and correspondingly Rewa district covers a total area of 6,240 km^2. Rewa (Luard 1907) is bounded by the districts of Mirzapur, Banda and Allahabad of Uttar Pradesh in the North and North-East; in the South by Sidhi district and it shares the boundary with Satna district in the West. Ground truth data confirms the existence of numerous new painted rock shelter sites in the region. Apart from previous studies and documentation of two hundred and fifty rock-shelters, the present research has brought to light more than forty new rock-shelters with in situ archaeological record in the region. The regions of South and South-Western Mirzapur along the North and North-Eastern parts of Rewa have been explored thoroughly, in order to document and catalogue various new archaeological sites. Sandstone outcrop is quite common in the area, although all the shelters are not painted. A propensity of painting activity is seen along the South, South-West and North, North-East boundary regions of Mirzapur and Rewa. While survey in the said part of Mirzapur has been exhaustive, only a small part of the sandstone belt of the Rewa district of M.P. (Madhya Pradesh) has been surveyed due to time, logistics, resources and other associated constraints. Field-work data and close observation of landscape modifications for three consecutive years have revealed several ground truth points in the Vindhyan sandstone belts of Mirzapur and Rewa districts. Field observation confirmed the fact that the adjoining district of Rewa, situated at the South-Western boundary of Mirzapur, also falls under the Vindhyan range and is full of archaeological sites having tremendous potential.

2.2 Satellite and GIS Datasets

This research work has been realised using multispectral data from one of the Indian satellites named as IRS-P6 (ResourceSat-1) having Advanced Wide Field Sensor (AWiFS). AWiFS has similar payload like LISS-3 and/or LISS-4 (Linear Image Self-Scanning 3 and 4). Although the spatial resolution in meter scale in ResourceSat-1 AWiFS, ResourceSat-1 LISS III and ResourceSat-1 LISS IV (Multi-spectral mode) vary from 56 (at nadir), 24 to 5.8 (at nadir) meters. The spectral bands for the said AWiFS sensors are bands 2, 3, 4 and 5. The topographic maps from the Survey of India, Kolkata; District Resource Maps (DRM) from the Geological Survey of India, Kolkata, road maps of Mirzapur and Rewa have all been utilised for this study. The GPS locations for all the shelters are implemented as ground truth data along with the digitized and geometrically corrected maps to classify the landscape of the rock art into several unique parts for supervised classification. The orthoimages gathered from satellites have been used to generate digital elevation model for the Mirzapur and Rewa districts. Ground control points,

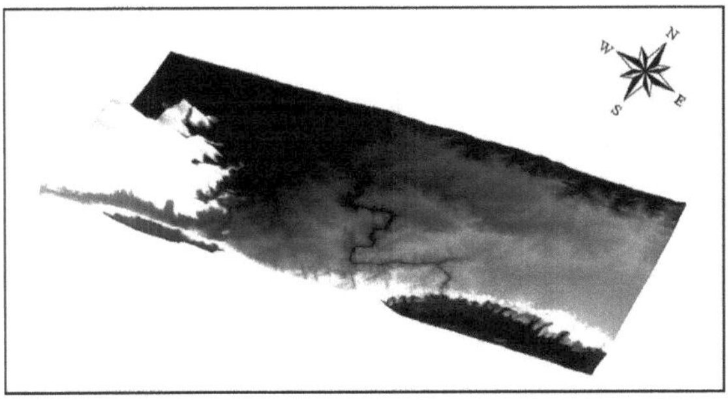

Fig. 2 Digital elevation model of the area

satellite ephemeris and radiometrically appropriated images have been included to measure the accuracy levels.

The visualisation of data, data acquisition, management, exploratory spatial data analysis (ESDA), Confirmatory data analysis (CDA) and interpretation in this work have been done following two different but unique techniques to map the local rock of the area that hosts a wealth of information. The applications of Bhuvan (http://bhuvan.nrsc.gov.in) have been implemented for the Cartosat1 and ResourceSat-1 images. Bhuvan is a geoportal of Indian Space Research Organisation (ISRO), Department of Space, Government of India (http://www.isro.org/). Additionally NNRMS (National Natural Resources Management System) data (http://www.nnrms.gov.in) is included in this work to consolidate the methodological rigour. Cartosat DEM data has proved to be very valuable for this work, refining and demonstrating the digital elevation model of the area (Fig. 2) in a 3D view. The ResourceSat-1 images have been processed, reclassified and georeferenced using ground truth data; obtained in the form of GPS points. The area is first divided into different small micro-areas depending on the GPS points coming from various parts of the region. Next, the whole region is divided into five broad classes namely; forest, waterbodies, sandstone, alluvial land and cropland respectively implementing the applications of Indian Remote Sensing satellite—P6 (ResourceSat-1). Each parameter has been colour coded uniquely to represent the variations in profile and to understand the total dataset. All the rock-shelter sites have been found distributed in a cluster or dispersed fashion within the sandstone complexes of the two said districts.

2.3 Classifiers/Algorithms Implemented in this Study

2.3.1 Artificial Neural Network (ANN)

Artificial neural network is used in various disciplines that basically work through the principles of simulation. The ANN model is based on three elementary layers namely, input or encoded, cryptic and output or linear layer that might jointly assume any logical function. Every neuron signifies an encoded function in digital image processing. If one neuron is described by one image band at the input layer; therefore every neuron in the linear layer is corroborated by a representative class. This model is independent of statistical interconnectivity and does not look for a normal distribution within the data. On the contrary it is very adaptive; relying on the estimates of a series of functions derived from the data. The neural network model will become increasingly important in archaeological remote sensing applications in mapping the surface of the prehistoric sites and probable archaeological regions. In recent years it has become imperative to classify the land cover and present day land use patterns of an area of rich cultural and/or archaeological importance. The images in the artificial neural network paradigm have been classified on a pixel to pixel basis (Rumelhart 1989) that delimits the problems of pixel mixing. The ANN parameters need to be optimised for a stable result and thus require a preliminary analysis using validation data. In this paper, the ANN algorithms are optimised by the method as described in (Varshney and Arora 2004; Srivastava et al. 2012b). The number of training datasets chosen for optimisation comprises of 50 pure pixels per class. However, for accurate estimation of optimised results, another set of 30 pure pixels per class has been taken into account for the validation purposes.

The exactitude of the geo-referenced image is verified with an existing geo-referenced map of the area; in order to ascertain uniformity of the data, in terms of maps, projection information and layer stacks. The ANN classifier divided the area into five classes, namely, forest, waterbodies, alluvial land, cropland and sandstone areas. In this ANN based image the parameter, sandstone is seen distributed proportionally in the area which adheres to the field data.

The neural network classifier used was a layered feed-forward model in ENVI (Environment for Visualizing Images) version 4.8 (ITT Visual Information Solutions SA) with standard back propagation for supervised learning. This category of ANN logistic is particularly superior for supervised classification because of its facility to learn by pattern and simplify the process (Srivastava et al. 2012b). In the case of ANN, the setting of the appropriate number of hidden layers depends upon the structure of the input data. Usually, it ranges from one to three. In practice, one layer of hidden nodes is sufficient in most cases (Gao 2009), hence a hidden layer of one is used in this study. The learning algorithm uses back propagation, which is one of the most commonly used forms of neural computing in remote sensing. The ANN weights were initialized using a uniform distribution. Learning rate was set to 100 for the hidden layer and 0.01 for the output layer, while stopping criteria is fixed to 0.001. The typical logistic activation function can be expressed as Eq. 1 (Schalkoff 1997; Friedman and Kandel 1999):

$$o_j = 1/(1 + e^{-\lambda net_j}) \tag{1}$$

where, o_j is the output of external input j, λ is a gain factor. The term net_j can be computed using Eq. 2 (Schalkoff 1997):

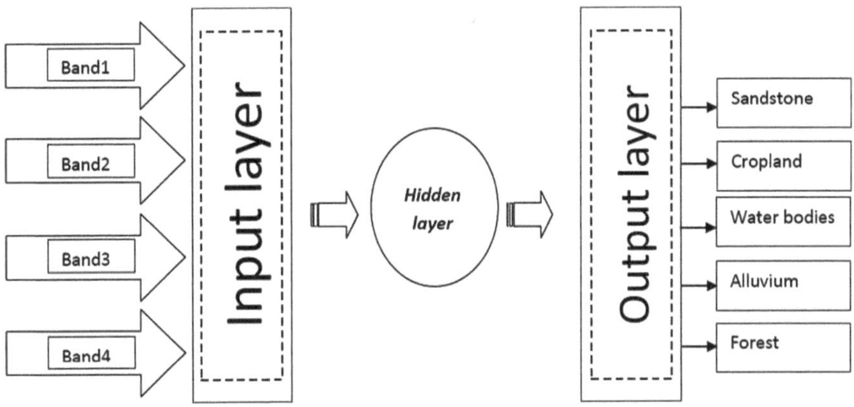

Fig. 3 Structure of artificial neural network

$$net_j = \sum_i w_{ji} o_i \tag{2}$$

where, w_{ji} is the weight of interconnection channel to unit j from unit i and o_i is the output of external unit i. The structure of ANN is represented in (Fig. 3).

2.3.2 Maximum Likelihood Classification (MLC)

The maximum likelihood classification is a statistical technique that has varied applications in remote sensing studies. This supervised classificatory scheme for remote sensing data works through the principles of probability statistics and relies on Bayes' theorem of decision making. MLC divided the entire region under study into several specific sub classes as training zones assuming a Gaussian normality. Each unique region is modified into pixels through the computer algorithm. These pixels are then represented as uniform spectral classes; that is denoted by the mean vector and covariance matrix. The pixels belonging to every class is then calculated using probability density function. The regions are classified depending on the similarity of each pixel cells with the other related ones, which is highly probabilistic and forms a unique class. The maximum probability counts of the pixels as individual cells get the highest value in order to attain a separate entity as a class within the entire land cover (Mustapha et al. 2010). The Maximum Likelihood Classification tool is considered for image classification, as it is incorporating both the variances and covariances of the class signatures and assigning each cell to one of the classes represented in the signature file. The algorithm used by the Maximum Likelihood Classification tool is based on Bayes' theorem and the equation used in MLC classification is shown in Eq. (3) (ERDAS 1999).

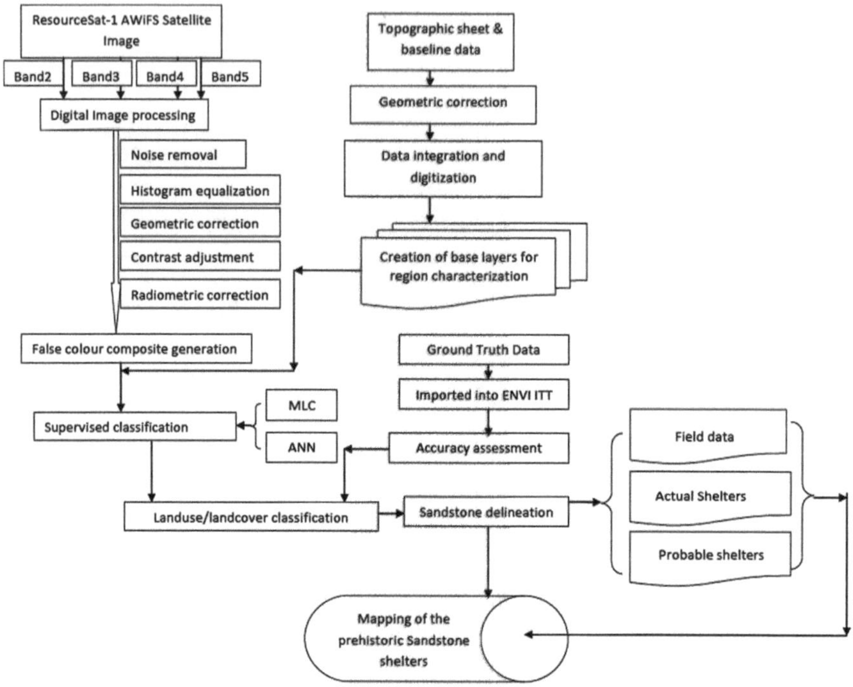

Fig. 4 Flowchart of the methodology used in the study

$$D = \ln(a_c) - [0.5\ln(|\text{cov}_c|)] - [0.5(\mathbf{X} - \mathbf{M}_c)\text{T}(\text{cov}_c - 1)(\mathbf{X} - \mathbf{M}_c)] \quad (3)$$

where, D is weighted distance; c is a particular class; **X** is the measurement vector of the particular pixel; $\mathbf{M}_c$ is the mean vector of the sample of class; a_c is percent probability that any particular pixel is a member of class c; (Defaults to 1.0); Cov_c is the covariance matrix of the pixels in the sample of class c; $|Cov_c|$ is determinant of Cov_c; Cov_c-1 is inverse of Cov_c; ln is natural logarithm function; T= transposition function. The MLC technique has successfully classified the documented area into forest, waterbodies, alluvial land, cropland and sandstone areas. The flowchart depicting the methodology used in this study is shown in (Fig. 4).

3 Accuracy Assessment of the Classified Images

In order to evaluate the performance of the algorithms, the accuracy assessment was carried out using the validation datasets collected during the field visits, assuring distribution in a sensible pattern so that a particular number of observations were assigned to each category on the classified image generated by the classifiers. The Kappa accuracy can be computed using equation (4) (Bishop et al. 1975).

$$\kappa = \frac{N \sum_{i=1}^{r} X_{ii} - \sum_{i=1}^{r} (x_{i+})(x_{+i})}{N^2 - \sum_{i=1}^{r} (x_{i+})(x_{+i})} \quad (4)$$

where, r is the number of rows in the matrix, X_{ii} is the number of observations in row i and column i (the diagonal elements), x_{+i} and x_{i+} are the marginal totals of row r and column i, respectively, and N is the number of observations. The total integrated accuracy, which has been measured according to defined proportion of the assessed area is classified correctly and can be estimated by Kappa statistics (Srivastava et al. 2012b). The user's accuracy calculated the proportion of pixels classified as belonging to a class that truly belongs to that class, while the producer's accuracy provides the proportion of pixels truly belong to a class that are classified as belonging to that class (Srivastava et al. 2010). The user's and producer's accuracies measurements are related to commission and omission errors (Congalton 1991; Mukherjee et al. 2009; Gupta and Srivastava 2010).

4 Results and Discussion

4.1 Land Covers Distribution and Accuracy Assessment

The classification maps produced from the implementation of the ANN and MLC are illustrated in (Fig. 5a, b). The highest accuracy is shown by ANN, hence only ANN classified image is taken into account for class distribution analysis. The LULC distributions and their percentage cover determined using ANN techniques is shown in (Table 1). In this study the IRS P6 ResourceSat-1 satellite images have been classified into five classes named as forest, waterbodies, sandstone, alluvial land and cropland. The classes created and the proportion of total area of the image covered by them, provide an insight to the composition of the total area (Townshend et al. 1991). On account of analysis of these classified images, it is possible to infer, to a certain extent, the changes that occurred in spatial composition of different physiographic features (Mukherjee et al. 2009). ANN classification is found to be the best by analysing the kappa statistics and accuracy. Overall ANN accuracy is 84.29 %, where ground truth data for the individual subsets vary. The IRS P6 ResourceSat-1 data is based on ANN classification, the values in terms of producer accuracies are obtained for the forest as 96.67 %, water bodies 96.3 %, alluvial land 82.35 %, cropland 91.89 % and sandstone 69.84 % respectively as separate classes, while the user accuracy obtained for the same classes in the similar order are 87.88, 92.86, 100, 61.82 and 93.62 % respectively. The kappa Coefficient for ANN classification is 0.80. On the contrary, the accuracy percent for the MLC based assessment is 81.15 %; where the Kappa Coefficient is 0.76. The percentage of the classes varies accordingly (Table 2). The producer accuracy obtained during

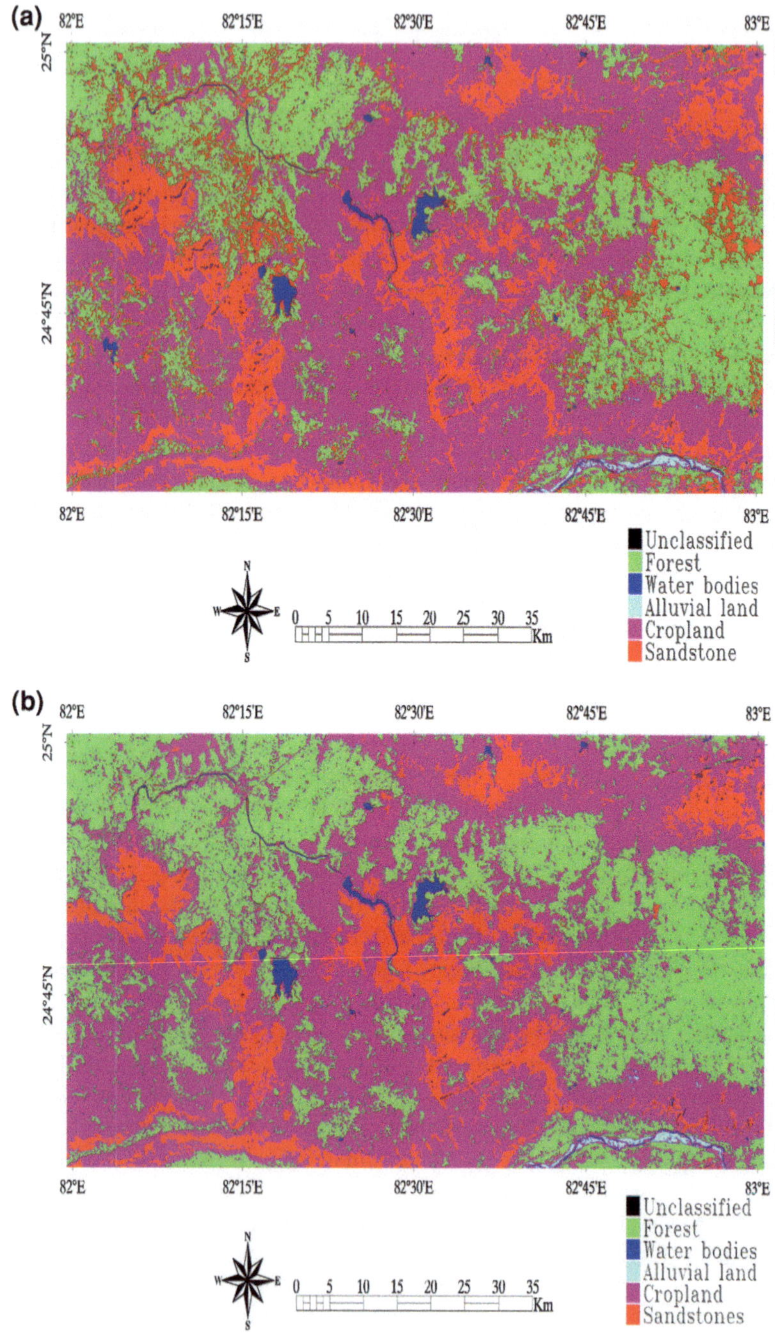

Fig. 5 Classified image of IRS P6 (ResourceSat-1) using ANN (**a**) and MLC (**b**)

Table 1 Land use/land cover distribution of the study area

Class name	Land cover (%)
Forest	29.58
Waterbodies	1.06
Alluvial land	0.27
Cropland	48.54
Sandstone	20.55
Total	100.00

Table 2 Brief results of the overall classification accuracies procured by applying the ANN and MLC methods on the IRS P6 ResourceSat-1 data

LULC	ANN		MLC	
	Producer accuracy (%)	User accuracy (%)	Producer accuracy (%)	User accuracy (%)
Forest	96.67	87.88	96.67	82.86
Waterbodies	96.30	92.86	85.19	92.00
Alluvial land	82.35	100.00	82.35	100.00
Cropland	91.89	61.82	89.19	57.89
Sandstone	69.84	93.62	66.67	91.30
Overall accuracy (%)	84.29		81.15	
Kappa coefficient	0.80		0.76	

MLC classification are Forest (96.67 %), waterbodies (85.19 %), alluvial land (82.35 %), cropland (89.19 %) and sandstone (66.67 %), while the user accuracy obtained for the similar class order are 82.8, 92.0, 100, 57.89 and 91.3 % respectively. This demonstrates the quality and strength of two different techniques where they are able to map the same landscape cover allotting different values and percentages to discrete elements. The overall area distribution based on ANN method assigns 20.55 % to sandstone, 48.54 % to cropland, 0.27 % to alluvial land, 1.06 % to waterbodies and finally 29.58 % to the forest cover. The classification of the archaeological landscape helps to characterize the parameters directly associated with the site formation processes (Binford 1981; Mol and Viles 2010; Grab et al. 2011). In this work, delineation of sandstone has got ultimate importance to map the rock-shelters in the surveyed and unexplored parts of the Mirzapur and Rewa districts.

4.2 *Interpretation of the Results and Archaeological Relevance*

The applications of two techniques in this study have been successfully able to map the distribution of sandstone in the field area and beyond it. Invariably one method produces relatively better results than the other one with a certain degree

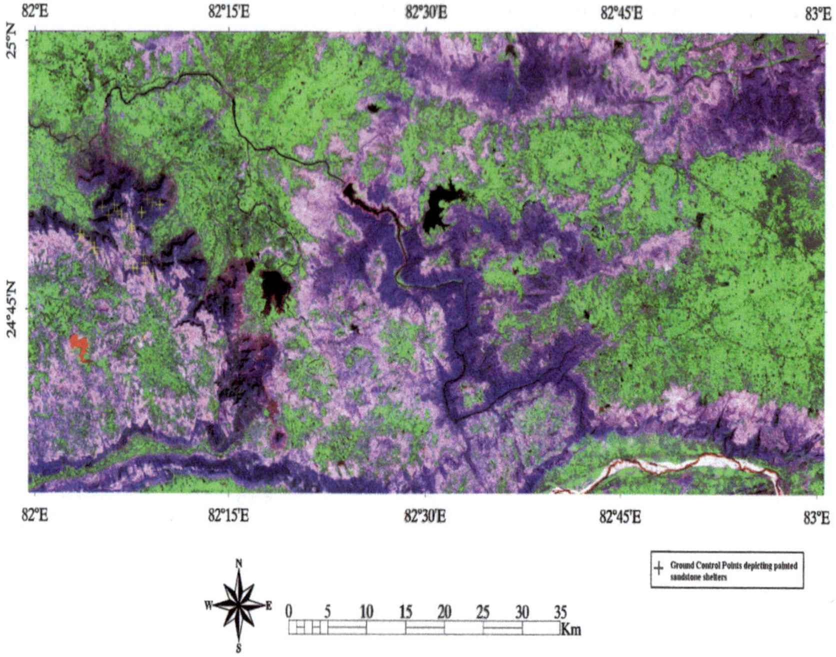

Fig. 6 Ground truth points collected during field survey

of accuracy. In this research particularly the investigation and mapping of sandstone within the landscape has been of primary importance, since all the documented painted rock-shelters are made of sandstone. Most importantly constructing the archaeological predictive map for sandstone as a locational tool for finding out new and potential painted shelters in central Indian context would open up new research paradigms. The number of sites and their preservation, conservation is directly related to the nature and patterning of sandstone. The South-Western part of Mirzapur has been exhaustively surveyed and more than forty new sites have been discovered. The region towards the North and South of Mirzapur district, where it shares border with Rewa also has very many painted rock-shelters, whereas a paltry of them could be catalogued in the field recently. The ANN and MLC both detected a linear trend of rock-shelters distributed across the landscape of Rewa. The present field-work data comes from the mid ridge of the elongated sandstone belt consisting of South-Western Mirzapur and parts of Rewa. The land cover in the later resembles the landscape of Mirzapur, revealing cropland, forest areas, waterbodies, alluvial land and finally sandstone. The accurate mapping of sandstone made it possible to pinpoint the sites within the proper context of the art, since GPS locations for all of the sites have been included in the data analyses (Fig. 6). The patterning of the sites show a tendency of propagation, where the sites are following the natural formation of sandstone in both directions, southwards and northwards from the borders of Mirzapur towards and across the Rewa district of Madhya Pradesh.

The practical and logical parameters set by the local microclimate at Mirzapur, occasionally represent environmentally and climatologically disturbed landscape. The distribution of painted sandstone shelters is indicative of the function of elevation and terrain morphology. Ground survey suggested that topography could be an important parameter which can affect the rock art of an area. In ancient times also, the area with very low slope and nearer to river might have had a greater chance of flood risk than higher one, which is here assumed that this might affect the hunter gatherers also in the past. So, it is presumed that the chances of getting rock art are more towards higher slope than lower ones. Hence, the DEM of the area has been utilized to delineate and understand the possible terrain features. It has been found that the frequency of the sites decreases gradually with a lower slope. The comparison with GPS points reveal that almost all the sites are situated in higher slope areas along the South-Western parts of the Mirzapur and adjoining Rewa district and thus supported the hypothesis outlined in the study. Characterising the present sites in terms of DEM and the classification results there is possibility of finding other new sites in the higher slope areas of the Mirzapur and the Rewa district. The models generated in this study along with the supposedly theoretical premises also support this observation. Enhanced density of rock-shelters are located at the higher elevations suggesting sacred or ritual sites hidden within the thickly vegetated forest cover and the presence of fewer sites at the lower elevations along the small rivulets suggesting camp sites where the prehistoric hunter-gatherers might have preferred to carry out their utilitarian activities.

A clear trend has been observed from the ground truth data and field-work that the rock-shelter sites are following the extension of the sandstone formation. The sandstone shelters documented at the South-Western part of the Mirzapur district have shown a gradual progression in two directions towards the adjoining Rewa district. Field data suggests that geologically Rewa district belongs to the Mesoproterozoic to late Neoproterozoic Vindhyan Supergroup revealing Rewa sandstone, shale, Kaimur and Bhander group of rocks (Tiwari 2011) and is bordered by the Mirzapur district of U.P. in the Northwest, North and Northeast. Field-data and/or ground truth data suggest that the total number of painted sites gradually increases towards the Rewa district following the distribution of the host-rock, which is sandstone. The paintings here ranged from the hunting gathering stage, depicting different types of animals (Fig. 7a) to the historic phase, where battle scenes predominate (Fig. 7b). The models generated in this work, by means of different techniques and satellite imageries therefore suggest, that further field-work along the sandstone formations might assist to detect numerous new painted shelters and other Mesolithic to late upper Palaeolithic, and ultimately sites of historic age consisting of archaeological artefacts (Fig. 7c). The sandstone in and around the areas of Mirzapur today is a subject of large scale quarrying. The recent knowledge on the relationship of sandstone and painted rock-shelters authenticated the beginning of a new research paradigm in Central Indian rock art and

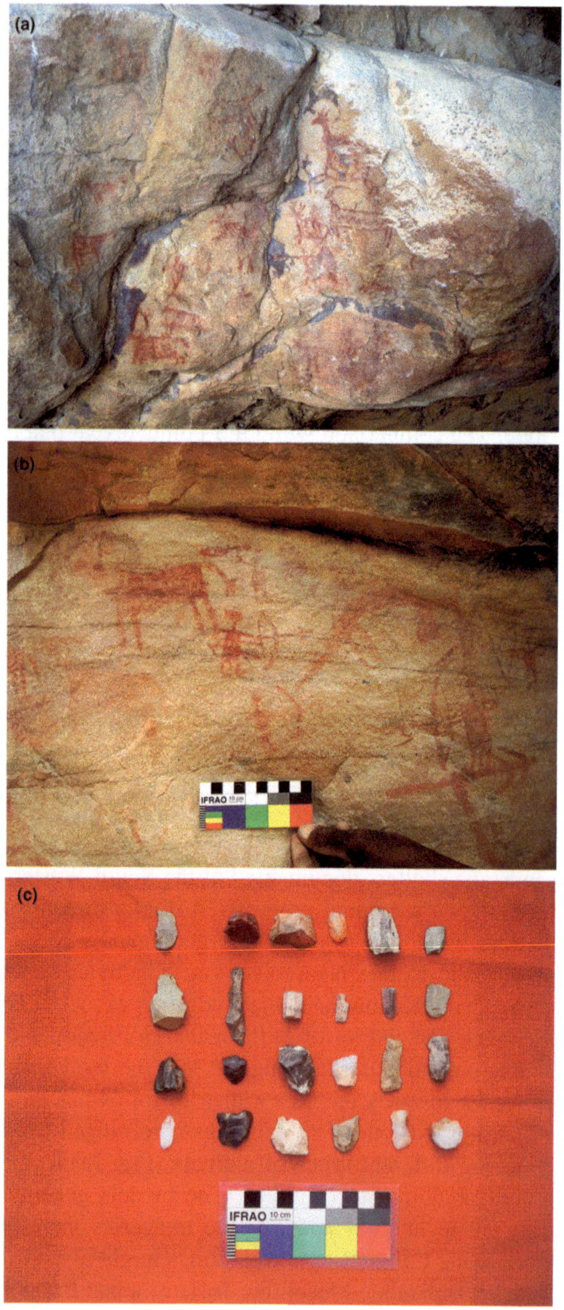

Fig. 7 **a–c** Rock shelter and Painted sites located on the landscape discovered during field survey (**a** Animals depicted in the rock art; **b** battle scene in the rock art; **c** Stone tools or archaeological artefacts found from the rock-shelters)

archaeology. Therefore, it is not widely known yet. The lack of specialist knowledge and neglect from the local population is leading to the destruction of several painted sites of intense heritage value in Mirzapur. This predictive model would help to determine new regions for pilot surveys to discover new sites. Apart from this, the models for locational and landscape archaeology would be beneficial to come up with the contingency plans to rescue and protect the sites, that are under the constant threat of destruction. Most of the times, painted shelters have been detected along with large amount of archaeological materials, present on the surface of the sites. The quarrying of sandstone, sometimes panels with prehistoric paintings of invaluable tangible heritage value in this region is gradually obliterating all the traces of prehistoric cultural heritage from the area. Therefore, accurate mapping of sandstone, along with the land cover and present day land use pattern in this region play an important role not only to locate new sites, but also to protect the global cultural heritage and local natural resources.

5 Conclusion

The accurate mapping of sandstone that hosts prehistoric art has been able to reveal a few important parameters in terms of the understanding of local geology, used and exploited by the prehistoric hunter-gatherers and technicalities involved in the process and methodologies of RS and GIS. The ground truthing and survey work together identified the distribution and patterning of rock-shelter sites in the Mirzapur and Rewa districts. Preliminary survey and cataloguing of the art in the region showed basic inter-site clustering and rudimentary patterns. The innumerable number of sites located in this region demonstrates the importance of rock in rock art and archaeology and that rock in itself is very important just as rock art. The movement of rock art, in terms of shelters scattered across the landscape of the art is defined and constructed by sandstone. The prehistoric hunter-gatherers understood the value of sandstone, not only to create art for communicative purposes but also to demarcate the landscape. The host-rock, which is sandstone, both represents prehistoric lifeways through long painting activities and local geology that remained partly unaltered in the region. The applications of ANN and MLC for the dataset on rock art suggests a higher accuracy level for the ANN subset and hence the classifiers. With a differential degree of accuracy, both ANN and MLC have been able to map the natural sandstone outcrop and the location of painted rock-shelters within the landscape characterising all other features; which indicates the viability for these techniques in landscape archaeology and archaeological locational modelling. This study has detected a clear combination of the co-existence of geological formation of sandstone and archaeological sites in the form of painted rock-shelters in the districts of Mirzapur and Rewa. The sandstone shelters followed a linear trend criss-crossing the South-Western border of the Mirzapur district and Eastern part of the Rewa district establishing the importance of local geology in archaeological site formations and the mobility of rock art in the region exposing the elements of hunter-gatherer ways of life. Finally, the present results on sandstone mapping suggest that future studies could be focussed on the adjoining and extended sandstone belts of the Mirzapur and Rewa districts in order to find several new archaeological sites with prehistoric and/or historic paintings and stone tools.

Acknowledgment The authors are extremely grateful to the Archaeological Survey of India, Janpath, New Delhi for the permission to commence field-work in the Mirzapur region of Central India. The first author extends his sincere thanks to the local guides and villagers of Mirzapur and Rewa districts for their cordial and unflinching help and hospitality during the field-work, exploring, documenting and cataloguing several new sites in this region of Central India. The first author received University of Bristol Centenary Research Scholarship which supported this research endeavour. The research travel grants provided by the Graduate School of Arts and Humanities, University of Bristol, are highly appreciated.

References

Allchin B, Allchin FR (1982) The rise of civilization in India and Pakistan. Cambridge University Press, Cambridge

Banerjee R, Srivastava PK (2013) Reconstruction of contested landscape: detecting land cover transformation hosting cultural heritage sites from central India using remote sensing. Land Use Policy 34(0):193–203. Available from: http://www.sciencedirect.com/science/article/pii/S0264837713000513

Binford LR (1981) Behavioral archaeology and the "pompeii premise". J Anthropol Res 37(3):195–208

Bishop YMM, Fienberg SE, Paul W (1975) Holland. Discrete multivariate analysis: theory and practice. MIT Press, Cambridge

Brown JA (1889) On some small highly specialized forms of stone implements, found in Asia, North Africa, and Europe. J Anthropol Inst GB Irel 18:134–139

Cockburn J (1894) On flint implements from the Kon ravines of South Mirzapur. J Asiat Soc Bengal Proc 63:21–37

Congalton RG (1991) A review of assessing the accuracy of classifications of remotely sensed data. Remote Sens Environ 37(1):35–46

Courty MA, Carbonell E, Vallverdú Poch J, Banerjee R (2012) Microstratigraphic and multi-analytical evidence for advanced neanderthal pyrotechnology at abric romani (capellades, spain). Quat Int 247:294–312

Diao Y, Ganesan D, Mathur G, Shenoy P (2007) Rethinking data management for storage-centric sensor networks. In: CIDR vol 7. pp 22–31

Dixon B, Candade N (2008) Multispectral landuse classification using neural networks and support vector machines: one or the other, or both? Int J Remote Sens 29(4):1185–1206

Drake-Brockman D (1911) Mirzapur: a gazetteer. Superintendent, Government Press, United Provinces

Erdas L (1999) Erdas field guide. Erdas Inc., Atlanta 698

Friedman M, Kandel A (1999) Introduction to pattern recognition: statistical, structural, neural, and fuzzy logic approaches. World Scientific Publishing Co. Inc., Singapore

Gao J (2009) Digital analysis of remotely sensed imagery. McGraw-Hill Professional, New York

Gasparini L, Manduchi R, Gottardi M, Petri D (2010) Performance analysis of a wireless camera network node. IEEE, pp 1331–1336

Grab SW, Goudie AS, Viles HA, Webb N (2011) Sandstone geomorphology of the golden gate highlands national park, South Africa, in a global context. Koedoe 53(1):01–14

Gupta M, Srivastava PK (2010) Integrating GIS and remote sensing for identification of groundwater potential zones in the hilly terrain of pavagarh, gujarat, india. Water Int 35(2):233–245
Jaiswal RK, Saxena R, Mukherjee S (1999) Application of remote sensing technology for land use/land cover change analysis. J Indian Soc Remote Sens 27(2):123–128
Jalan S, Sokhi BS (2012) Comparison of different pan-sharpening methods for spectral characteristic preservation: multi-temporal CARTOSAT-1 and IRS-P6 LISS-IV imagery. Int J Remote Sens 33(18):5629–5643
Kandrika S, Roy P (2008) Land use land cover classification of orissa using multi-temporal irs-p6 awifs data: a decision tree approach. Int J Appl Earth Obs Geoinf 10(2):186–193
Kavzoglu T, Colkesen I (2009) A kernel functions analysis for support vector machines for land cover classification. Int J Appl Earth Obs Geoinf 11(5):352–359
Lasaponara R, Masini N (2012) Remote sensing in archaeology: From visual data interpretation to digital data manipulation. In: Satellite Remote Sensing. Springer, Netherlands, pp 3–16
Luard C (1907) Rewa state gazetteer, vol IV. (Text and Tables), p 90
Misra V (2001) Prehistoric human colonization of India. J Biosci 26(4):491–531
Mol L, Viles H (2010) Geoelectric investigations into sandstone moisture regimes: Implications for rock weathering and the deterioration of san rock art in the golden gate reserve, South Africa. Geomorphology 118(3):280–287
Mukherjee S (2004) Textbook of environmental remote sensing. Macmillan, New York
Mukherjee S, Shashtri S, Singh C, Srivastava P, Gupta M (2009) Effect of canal on land use/land cover using remote sensing and GIS. J Indian Soc Remote Sens 37(3):527–537
Mustapha M, Lim H, Jafri MZM (2010) Comparison of neural network and maximum likelihood approaches in image classification. J Appl Sci 10(22):2847–2854 (Faisalabad)
Oommen T, Misra D, Twarakavi NKC, Prakash A, Sahoo B, Bandopadhyay S (2008) An objective analysis of support vector machine based classification for remote sensing. Math Geosci 40(4):409–424
Pappu S, Akhilesh K, Ravindranath S, Raj U (2010) Applications of satellite remote sensing for research and heritage management in indian prehistory. J Archaeol Sci 37(9):2316–2331
Pappu, S, Akhilesh K., Ravindranath S, Raj U, Gunnell Y (2010b) Research and management of palaeolithic resources in Tamil Nadu, South India. Antiquity 84(325):1596–1599
Pratap A, Kumar N (2009) Painted rockshelters at Wyndham falls, Mirzapur, Uttar Pradesh, India. Antiquity 83(321):891
Punia M, Joshi P, Porwal M (2011) Decision tree classification of land use land cover for Delhi, India using IRS-P6 AWiFS data. Expert Syst Appl 38(5):5577–5583
Rumelhart DE (1989) The architecture of mind: a connectionist approach. In: Posner MI (ed.). Foundations of cognitive science. MIT Press, Cambridge, pp 133–159
Schalkoff RJ (1997) Artificial neural networks. McGraw-Hill Companies, New York
Shukla A, Arora M, Gupta R (2010) Synergistic approach for mapping debris-covered glaciers using optical–thermal remote sensing data with inputs from geomorphometric parameters. Remote Sens Environ 114(7):1378–1387
Srivastava PK, Gupta M, Mukherjee S (2012a) Mapping spatial distribution of pollutants in groundwater of a tropical area of india using remote sensing and gis. Appl Geomatics 4(1):21–32. Available from: http://dx.doi.org/10.1007/s12518-011-0072-y
Srivastava PK, Han D, Rico-Ramirez MA, Bray M, Islam T (2012b) Selection of classification techniques for land use/land cover change investigation. Adv Space Res 50(9):1250–1265. Available from: http://www.sciencedirect.com/science/article/pii/S0273117712004218
Srivastava PK, Mukherjee S, Gupta M (2010) Impact of urbanization on land use/land cover change using remote sensing and GIS: a case study. Int J Ecol Econ Stat 18(S10):106–117
Tiwari R (2011) Assessment of groundwater quality and pollution potential of Jawa block Rewa district, Madhya Pradesh, India. Proc Int Acad Ecol Environ Sci 1(3–4):202–212
Tiwari SK (2000) Riddles of Indian rockshelter paintings. Sarup & Sons, New Delhi

Townshend J, Justice C, Li W, Gurney C, Mcmanus J (1991) Global land cover classification by remote sensing: present capabilities and future possibilities. Remote Sens Environ 35(2):243–255

Varshney PK, Arora MK (2004) Advanced image processing techniques for remotely sensed hyperspectral data. Springer, Berlin

Yu L, Zhang D, Holden EJ (2008) A fast and fully automatic registration approach based on point features for multi-source remote-sensing images. Comput Geosci 34(7):838–848

MIX
Papier aus verantwortungsvollen Quellen
Paper from responsible sources
FSC® C105338

If you have any concerns about our products,
you can contact us on
ProductSafety@springernature.com

In case Publisher is established outside the EU,
the EU authorized representative is:
Springer Nature Customer Service Center GmbH
Europaplatz 3, 69115 Heidelberg, Germany

Printed by Libri Plureos GmbH
in Hamburg, Germany